ALLE ZEIT WACH
SJ
1842

Fluid Dynamics as a Localizing Factor for Atherosclerosis

The Proceedings of a Symposium
Held at Heidelberg, FRG, June 18–20, 1982

Editors: G. Schettler R. M. Nerem
H. Schmid-Schönbein H. Mörl C. Diehm

With 90 Figures

Springer-Verlag
Berlin Heidelberg New York Tokyo 1983

Professor Dr. Dr. h.c. mult. Gotthard Schettler
Medizinische Klinik (Ludolf-Krehl-Klinik) der Universität
Bergheimer Straße 58, D-6900 Heidelberg 1

Dr. Robert M. Nerem
Department of Mechanical Engineering, University of Houston
Houston, TX 77004, USA

Professor Dr. Holger Schmid-Schönbein
Abteilung Physiologie, Medizinische Fakultät der
Rheinisch-Westfälischen Technischen Hochschule
Schneebergweg 211, D-5100 Aachen

Professor Dr. Hubert Mörl
Medizinische Klinik (Ludolf-Krehl-Klinik) der Universität
Bergheimer Straße 58, D-6900 Heidelberg 1

Dr. Curt Diehm
Medizinische Klink (Ludolf-Krehl-Klinik) der Universität
Bergheimer Straße 58, D-6900 Heidelberg 1

ISBN-13: 978-3-642-69087-7 e-ISBN-13: 978-3-642-69085-3
DOI: 10.1007/978-3-642-69085-3

Library of Congress Cataloging in Publication Data. Main entry under title: Fluid dynamics as a localizing factor for atherosclerosis. Bibliography: p. Includes index. 1. Atherosclerosis-Etiology-Congresses. 2. Hemodynamics-Congresses. I. Schettler, G. (Gotthard), 1917-. RC692.F55 1983 616.1'36071'015321 83-14715

Printing: Beltz Offsetdruck, 6944 Hemsbach/Bergstr. Binding: J. Schäffer OHG, 6718 Grünstadt
2121/3140-543210

Preface

This volume represents the proceedings of a satellite meeting of the VIth International Symposium on Atherosclerosis entitled "Fluid Dynamics as a Localizing Factor for Atherosclerosis". Contained within are the manuscripts of the technical presentations offered at the meeting together with the discussion recorded at that time, which was then edited and abbreviated for inclusion in this volume.

The meeting was held in Heidelberg, Germany, on June 18-20, 1982. Though the first of its type on this specific subject held in Europe, it is in fact an outgrowth and a continuation of a series of meetings held in the United States. The first of these was held at the Ohio State University in Columbus, Ohio, in 1974, the second at Delaware Water Gap, Pennsylvania, in 1976, the third again at Ohio State University in 1978, and the fourth and most recent at the University of Houston, Texas, in 1980. In these the aim has been to bring together a small interdisciplinary group of clinicians, engineers, life scientists, and theoreticians in an atmosphere which would facilitate the open exchange of ideas, methodologies, and the most recent of results. Whereas the first meeting in 1974 was focused largely on fluid dynamics as related to the pathogenesis of arterial disease, there has been a gradual, but noticeable shift to an emphasis on the role of fluid mechanics in the biology of the arterial wall.

Over the years, it has become evident that the biology of the arterial wall cannot be understood unless details of intraarterial and transmural flow are included in the conceptual framework. The meeting therefore not only reflected a more international, but also a more interdisciplinary discussion of the topic of atherogenesis, a discussion which embraced the variety of conceptual approaches necessary to understand the physiology and pathophysiology of the arterial wall. In this way, the putative role of fluid dynamic factors in the process of atherogenesis can be put into the context of the other biochemical and biophysical processes that influence the slow transition from the normal to the diseased state.

This meeting would not have been possible without the support of the National Science Foundation, which funded the participation of the U.S. delegation as part of the U.S.-German scientific exchange program, and the Thyssen Foundation, Cologne, which supported the participation of the German scientists, as well as those from other countries. Additional support was provided by Pharma-Schwarz, Monheim; Deutsche Abbott GmbH, Wiesbaden; and Albert Roussel, Wiesbaden.

The meeting was held in the excellent conference facilities of the Max-Planck-Haus in Heidelberg. It is also a pleasure to acknowledge the hospitality and administrative assistance provided by the members of the Department of Internal Medicine, University of Heidelberg. We are particularly indebted to Ms. Annerose Hack and Ms. Klara R. Jurutka.

Also, the participants are to be thanked for the spirit in which they entered into the discussions which took place.

Finally, just prior to the publication of this proceedings, Professor Alexander Naumann from Aachen passed away unexpectedly. He was an active participant in the specialists meeting just nine months earlier, and his stimulating new approach to the problems of the meeting very much contributed to the success of the meeting.

G. Schettler
R.M. Nerem
H. Schmid-Schönbein
H. Mörl
C. Diehm

Contents

List of Formal Participants

Blackshear, P.L., Dr.
Department of Mechanical Engineering, University of Minnesota, Minneapolis, MN 55455, USA

Blasberg, P., Dr.
Abteilung Physiologie, Medizinische Fakultät der Rheinisch-Westfälischen Technischen Hochschule, Schneebergweg 211, D-5100 Aachen

Born, G.V.R., Professor Dr.
University of London, King's College, Department of Pharmacology, Strand, London WC2R 2LS, England

Buss, H., Professor Dr.
Pathologisches Institut, Friedrich Paffrath-Straße 100, D-2940 Wilhelmshaven

Caro, G., Dr.
Physiological Flow Studies Unit, Imperial College of Science and Technology, Prince Consort Road, London SW 7, England

Cornhill, J.F., Dr.
The Ohio State University, College of Medicine, Room 3168, Graves Hall, 333 West, 10th Avenua, Columbus, OH 43210, USA

Dewey, C.F., Jr., Dr.
Fluid Mechanics Laboratory, Department of Mechanical Engineering, Massachusetts Institute of Technology, Cambridge, MA 02139, USA

Diehm, C., Dr.
Medizinische Klinik (Ludolf-Krehl-Klinik) der Universität, Bergheimer Straße 58, D-6900 Heidelberg 1

Freudenberg, N., Dr.
Pathologisches Institut der Universität, Ludwig-Aschoff-Haus, Albertstraße 19, D-7800 Freiburg i. Br.

Friedman, M.H., Dr.
Applied Physics Laboratory, The Johns Hopkins University, Johns Hopkins Road, Laurel, MD 20707, USA

Giddens, D.P., Dr.
School of Aerospace Engineering, Georgia Institute of Technology, Atlanta, GA 30332, USA

Glagov, S., Dr.
Department of Pathology, The University of Chicago, School of Medicine, Chicago, IL 60637, USA

Jellinek, H., Professor Dr.
2nd Department of Pathology, Semmelweis Medical University, H-1091 Budapest, Hungary

Karino, T., Dr.
University Medical Clinic, Montreal General Hospital, 1650 Cedar Avenue, Montreal, Quebec H3G 1A4, Canada

Lea, George K., Dr.
Directorate of Engineering, National Science Foundation, Washington, DC 20550, USA

Liepsch, D., Professor Dr.
Fachhochschule München, Fachbereich 01, Versorgungstechnik, Postfach 20 09 27, D-8000 München 2

Meairs, S. Dr.
Anatomisches Institut III der Universität, Im Neuenheimer Feld 307, D-6900 Heidelberg 1

Mörl, H., Professor Dr.
Medizinische Klinik (Ludolf-Krehl-Klinik) der Universität, Bergheimer Straße 58, D-6900 Heidelberg

Naumann, A., Professor Dr. Dr. h.c.A.
Aerodynamisches Institut der Rheinisch-Estfälischen Technischen Hochschule, Templergraben 55, D-5100 Aachen

Nerem, R.M., Dr.
Physiological Fluid Mechanics Laboratory, Department of Mechanical Engineering, University of Houston, Houston, TX 77004, USA

Schettler, G., Professor Dr. Dr. h.c.mult.
Medizinische Klinik (Ludolf-Krehl-Klinik) der Universität, Bergheimer Straße 58, D-6900 Heidelberg 1

Schmid-Schönbein, H., Professor Dr.
Abteilung Physiologie, Medizinische Fakultät der Rheinisch-Westfälischen Technischen Hochschule, Schneebergweg 211, D-5100 Aachen

Schwartz, J.C., Dr.
Department of Pathology, Gradnate School of Biomedical Sciences, The University of Texas Health Science Center at San Antonio, 7703 Floyd Curl Drive, San Antonio, TX 78284, USA

Schwartz, S.M., Dr.
Department of Pathology SM-30, University of Washington, Seattle, WA 98195, USA

Seed, W.A., Dr.
Department of Medicine, Charing Cross Hospital Medical School (Fulham), Fulham Palace Road, London W 6 8RF, England

Spence, J.D., Professor Dr.
University of Western Ontario, Department of Medicine, Clinical Neurological Sciences and Pharmacology, Victoria Hospital, 375 South Street, London, Ontario N6A 4G5, Canada

Stehbens, W.E., Professor Dr.
Department of Pathology, Wellington Clinical School of Medicine, Wellington Hospital, Wellington 2, New Zealand

Walløe, L., Professor Dr.
Institute of Informatics and Physiology, University of Oslo, P.O. Box 1080 Blindern, Oslo 3, Norway

Weinbaum, S., Dr.
School of Engineering, The City College of the City University of New York, 140th Street and Convent Avenua, New York, NY 10031, USA

Wurzinger, L.T., Dr.
Abteilung Physiologie, Medizinische Fakultät der Rheinisch-Westfälischen Technischen Hochschule, Schneebergweg 211, D-5100 Aachen

Opening Comments

G. Schettler

The development of atherosclerosis is determined by two pathogenetic factors: the arterial wall and the contents of the vessel. Both factors should be considered as pathogenetic equals. This fact has not always been clear, and until recently the morphologically oriented atherosclerosis research had forgotten the importance of fluid dynamics? However, fluid dynamics played a large role in the development of the first pathogenetic concept. The German patho-anatomists Thomas and Rickert were closely related to this theory. Wilhelm Doerr undertook remarkable tests measuring and localizing the effect of dynamic force on the development of atherosclerosis. These experiments helped us to understand the morphology and topography of human atherosclerosis. From the field of physical chemistry came the discovery of the interfacial phenomenon which has opened up new areas of pathogenetic research. Included is the interaction between blood flow and the vessel wall, and this is a determining factor in the development of the primary lesion. From this area many more important findings can be expected.

This Symposium follows the VI International Atherosclerosis Symposium in Berlin. The program of the Berlin conference was extraordinarily broad. Themes ranged from pure morphology to clinical medicine, from prevention to rehabilitation, and from public health questions regarding atherosclerosis to special diagnostic and therapeutic techniques including surgery. Thus, an attempt was made in Berlin to include the entire palette of atherosclerosis research in the world.

This has resulted in the need for special meetings to bring specialists together in focused areas. Our satellite symposium here in the Max-Planck-Haus is just one example. As seen from the American-German cooperative work, research is international, and methods and experimental approaches are similar throughout the world. This is also true of our common goal, for structure and function are inseparably bound in medicine - the truth of which is shown in the living example of atherosclerosis.

Clearly animal experiments in this field must be interpreted with caution. Most experimental atherosclerosis models are not applicable to man. Only recently have comparable and even species identical experimental models been found. Our symposium thus will show that there is still much to be done and many questions to be answered. But this is the general principle of research, that new questions are always developing.

To the Greeks, a symposium means a discussion during and after a feast. Modern symposiums have not maintained this tradition. However, our gregarious organization will attempt to increase personal contact out-

Fluid Dynamics as a Localizing Factor for Atherosclerosis
Editors: G. Schettler et al.

side of this scientific sessions for the good of all our research goals. Therefore we are happy that this symposium has been so carefully planned with our American guests and that we as hosts can help make it a success. On this note I now officially open our meeting and wish that it will be a success for all of us.

Fluid Dynamic Approaches to Atherosclerosis

W. E. Stehbens

A. Introduction

The majority of investigators in the field of atherogenesis are not well versed in either the pathology of atherosclerosis or the natural history of the disease, and have accepted without question the validity of the lipid hypothesis, or some variation of it. There are those who, still believing that lipid is the sine qua non of atherosclerosis, rely on macroscopic sudanophilic (lipophilic) staining and on random histological sections to localize allegedly early lesions, which they investigate even in aortas of adults.

Nevertheless, there has been progress, for a most important advance of recent years has seen a belated but wide acceptance that arterial lesions, whether as spontaneously atherosclerotic in nature or as dietary-induced lipid deposits arise, mostly if not always, from areas of pre-existing intimal proliferation (Scott et al. 1979). Investigators are thus focussing attention on early intimal proliferation, the significance of which is crucial to any understanding of atherogenesis.

B. Intimal Proliferation

The intimal thickening present in the aorta and at the forks of distributing arteries in neonates and infants has often been regarded as physiological, i.e., as an integral part of the vessel wall, because it occurs in all infants, and because of a reluctance to accept that a pathological change could be universal at such an early stage (Stehbens 1979). However, the presence of smooth muscle in the intimal proliferation indicates that the role of the pads or cushions is not physiological because (i) the muscle is longitudinally orientated, (ii) there is evidence of cellular degeneration and the elastic tissue exhibits degenerative changes (Stehbens 1975c), and (iii) there is no evidence of a nerve supply to the thickenings, and their localization precludes any sphincteric effect (Stehbens 1979).

Another assumption is that intimal changes represent an adaptation to changing hemodynamics and an increase in vascular caliber during growth and maturation. Yet intimal thickening does not cease at maturation but continues throughout life, merging imperceptibly with overt atherosclerosis, and varying in degree from vascular bed to vascular bed, and from subject to subject. An appraisal of the significance of this intimal thickening involves detailed study of its localization and nature throughout life. Investigation has revealed that the intimal proliferation at arterial forks does not occur as a diffuse thickening, but as discrete intimal pads or cushions at specific anatomical sites about forks (Stehbens 1960, 1963a, 1979). These intimal thickenings at arterial forks are related to the flow divider and the region at the lateral angle where boundary layer se-

Fluid Dynamics as a Localizing Factor for Atherosclerosis
Editors: G. Schettler et al.

paration is likely (Stehbens 1975a). It has been demonstrated that there is a significant increase in the cross-sectional area of the stem prior to the actual division of the arterial lumen, and this expansion resembles a lateral or asymmetrical diffuser where boundary layer separation beyond the lateral angles can occur (Stehbens 1974a). The flow divider seems to act as a wedge, initiating vortex shedding, as in the jet edge effect. This appears to account for the intimal thickening in that region, and also for the physiological murmurs emanating from the vascular system (Stehbens 1975a). It has been postulated that the pressure oscillations in these regions evoked by the flow disturbances are responsible for the pathological changes in the wall at these particular sites. The possible dampening effect of erythrocytes, and the presence or absence of a marginal plasma zone in arteries, constitute research areas in need of future development.

Intimal thickening cannot be satisfactorily differentiated from overt atherosclerosis, with which it merges imperceptibly, and should be regarded as a consistent, integral stage of the disease. Analysis reveals that its distribution follows closely the distribution and varying severity of atherosclerosis in nature (Stehbens 1979). It has been demonstrated that intimal thickenings at arterial forks are sites of predilection for spontaneous and dietary-induced lipid deposition (Stehbens 1963b, 1964), and also regions where atherosclerosis subsequently develops in man. Their localization and development appear to be intimately related to hemodynamic stresses at the fork during physiological flow. More diffuse atherosclerotic changes may be the consequence of the distal extension of flow disturbances at forks, together with the effect of pulsatile flow and pressure augmented in the regions of the circulation where there is high mural stress.

C. Localization of Atherosclerosis

The other observations in pathology (Stehbens 1979) which indicate that not only the localization but also the severity of atherosclerosis is governed by hemodynamic factors are:

1. Augmented severity where the pulse and systolic pressures are high in the aorta and iliac arteries
2. Augmented severity at branching sites, junctions, curvatures and fusiform dilatations
3. The severity is directly related to the caliber of the vessel
4. Severity in the systemic, pulmonary and venous circulations is dependent on blood pressure
5. Hypertension in any system augments the severity, and hypotension indicates the likelihood of longevity
6. There is varying severity from vascular bed to vascular bed within the individual
7. Atherosclerosis in limb arteries is augmented by usage and diminished in a paralysed limb (MacCallum 1922; Moschcowitz 1929)
8. Atherosclerosis appears to run an accelerated course in berry aneurysms (Stehbens 1972)
9. Atherosclerosis affects all coats of the vessel wall, but involves the intima most severely, and the adventitia least severely
10. Predilection of the homolateral iliac artery to atherosclerosis in children with a single umbilical artery (Meyer and Lind 1974)

D. Hemodynamics as a Localizing Factor

Whilst such observations indicate the importance of hemodynamic factors in the pathogenesis of atherosclerosis, most authors still consider hemodynamics merely as a localizing factor in atherogenesis.

They have supported their hypothesis by experiments involving various types of injury to the arterial wall in animals subsequently subjected to dietary-induced hypercholesterolemia. The traumatized area is thus predisposed to the accumulation of lipid, and the deduction is then made that hemodynamic injury localizes atherosclerosis. This hypothesis is incorrect for two reasons. Firstly, cholesterol feeding may induce lipid storage in the vessel wall and in extravascular tissues, but it does not produce atherosclerosis or its complications (Stehbens 1979). There are irreconcilable differences between dietary-induced cholesterolosis of blood vessels and spontaneous or hemodynamically-induced atherosclerotic lesions (Stehbens 1979, 1981a), and cholesterol feeding does not produce aneurysms, intimal tears, ulceration or mural thrombosis. Secondly, in the hypercholesterolemic rabbit a recent arteriotomy wound accumulates lipid, but a recent phlebotomy wound in a vein, which is comprised of similar tissues though its architecture differs, is not predisposed to lipid accumulation (Stehbens 1975b). The correct conclusion to draw is that the injury, plus arterial hemodynamics perhaps acting synergistically, predisposes the artery to the lipid deposition, and not the injury per se, Furthermore, if in the same animal the contralateral artery and vein are anastomosed, the vein becomes the site of intense hemodynamic stress and accumulates more lipid than the afferent artery, or the control vein and arteriotomized artery (Stehbens 1973). Hemodynamics apparently governs the lipid deposition in the cholesterol-fed rabbit. Moreover, histological examination of the anastomosed vein reveals that the lipid deposition in such animals is a foam cell lesion, and this differs morphologically from true atherosclerosis, so that hemodynamic stress in the presence of dietary-induced hyperlipidemia does not necessarily produce the appearance of spontaneous lipid deposition of atherosclerosis. Such experiments do not preclude the possibility that hyperlipidemia in man may augment the spontaneous lipid deposition of true atherosclerosis.

E. Complications of Atherosclerosis

It has recently been reported that the complications of atherosclerosis (Fig. 1) may be explained on the basis of loss of tensile strength (Stehbens 1979), and histologically and ultrastructurally changes occur in the vessel wall that could be overt manifestations of a loss of cohesion of the vascular tissues, viz. medial thinning, elastic

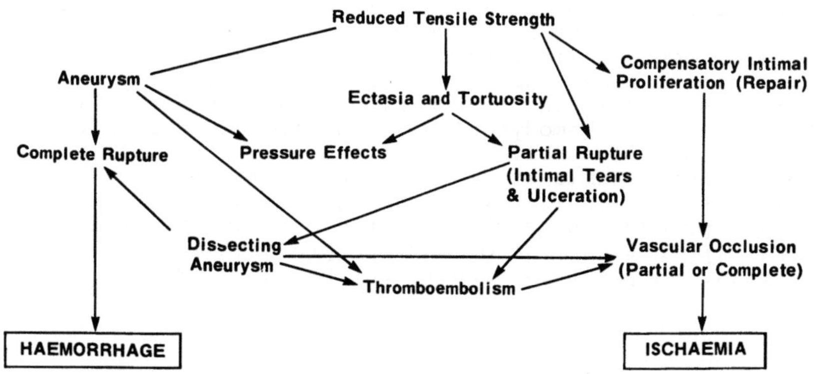

Fig. 1. Diagram illustrating the relationship of the complications of atherosclerosis to the loss of tensile strength of the vessel wall

tissue loss and fragmentation, cellular debris and separation of basement membranes from endothelium and smooth muscle cells (Stehbens 1975c). Stehbens (1958) postulated that atherosclerosis is due to hemodynamically-induced structural fatigue or failure of the vessel wall. He regarded atherosclerosis as the degenerative and reparative changes in the blood vessel walls that are the direct consequence of hemodynamically-induced fatigue (Stehbens 1979). To test this hypothesis, there was need to increase the vibrational stress on the vessel wall in order to accelerate the development of atherosclerosis and, consequently, two experimental models were used (viz. arteriovenous fistulae and aneurysms).

F. Hemodynamic Induction of Atherosclerosis

The creation of arteriovenous (carotid-jugular) fistulae in sheep causes the anastomosed vein to vibrate quite vigorously. Within four years the vein undergoes aneurysmal dilatation, gross thickening of the wall, intimal tears, mural dissection, thrombosis, wrinkling, hyalinized fibrotic areas, calcification and lipid deposition, all features of the atherosclerotic process (Stehbens 1974b). Ultrastructurally, the changes in the vein were also similar to human atherosclerosis (Stehbens 1975d). Control vessels on the contralateral side showed minimum phlebosclerosis. Similar atherosclerotic changes with complications have been observed in arteriovenous fistulae fashioned in patients with renal failure for hemodialysis, and these are therefore the human counterpart for the sheep experiments (Stehbens and Karmody 1975). More recently saccular, fusiform and berry aneurysms have been produced by microvascular surgery using autogenous vein grafts in normocholesterolemic rabbits. In the course of time, phlebosclerotic changes occur in the aneurysms and are followed by overt atherosclerosis with lipid deposition, calcification and mural thrombosis (Stehbens 1981b,c)). Feeding cholesterol to rabbits with experimental aneurysms does not accentuate the appearance of atherosclerosis but, on the contrary, produces a profound xanthomatous appearance of the aneurysmal walls, quite unlike the spontaneous lipid deposition of rabbits or humans (Stehbens 1981d). If hemodynamics is merely a localizing factor, then the cholesterol-feeding of rabbits with experimental arteriovenous fistulae or aneurysms should produce atherosclerotic changes, but the xanthomatous nature of the lipid deposition in these experiments (Stehbens 1973, 1981d) is unmistakable and argues against the hypothesis. On the other hand, the production of lipid deposition identical to that in man, as well as the complications of atherosclerosis in normolipidemic rabbits in which venous tissue, often regarded as being relatively immune to the disease, is subjected to intense hemodynamic stress, strongly supports the hypothesis that hemodynamics plays the major role in the etiology and pathogenesis of atherosclerosis and its complications (Stehbens 1974, 1981b,c). The hypothesis of hemodynamically-induced fatigue is the only theory current that plausibly explains the complications of atherosclerosis. If hemodynamic stress is regarded merely as a localizing factor, the assumption is that some major causative factor responsible for its production exists. But the evidence for the existence of some hypothetical factor must be produced and, in its absence, unscientific suppositions should not be put forward.

G. Conclusion

These experiments indicate the real likelihood that valuable information will be elicited from future research in hemorheological pathology. It is certain to contribute substantially to the field of atherosclerosis because blood vessels are not immune to the laws of hydrodynamics. Moreover, the experiments discussed indicated not only the susceptibility of the blood vessel wall to hemodynamic stress, but also that hemodynamic stress not only localizes but can induce atherosclerosis, irrespective of the serum lipid levels, and can produce it in both sheep and rabbits when on a 'non-atherogenic' diet.

Acknowledgements. Financial support from the Medical Research Council, the National Heart Foundation and the Neurological Research Foundation of New Zealand is gratefully acknowledged.

References

MacCallum WG (1922) Arteriosclerosis. Physiol Rev 2: 70-91

Meyer WW, Lind J (1974) Iliac arteries in children with a single umbilical artery: structure, calcifications and early atherosclerotic lesions. Arch Dis Child 49: 671-679

Moschcowitz E (1929) The cause of arteriosclerosis. Am J Med Sci 178: 244-267

Scott RF, Thomas WA, Lee WM, Reiner JM, Florentin RA (1979) Distribution of intimal smooth muscle cell masses and their relationship to early atherosclerosis in the abdominal aortas of young swine. Atherosclerosis 34: 291-301

Stehbens WE (1958) Intracranial arterial aneurysms and atherosclerosis. Thesis, University of Sydney

Stehbens WE (1960) Focal intimal proliferation in the cerebral arteries. Am J Path 36: 289-310

Stehbens WE (1963a) Histopathology of cerebral aneurysms. Arch Neurol 8: 272-285

Stehbens WE (1963b) The renal artery in normal and cholesterol-fed rabbits. Am J Path 43: 969-985

Stehbens WE (1964) Localization of spontaneous lipid deposition in the cerebral arteries of sheep. Nature 203: 1294-1295

Stehbens WE (1972) Pathology of the cerebral blood vessels. C.V. Mosby, St. Louis

Stehbens WE (1973) Experimental arteriovenous fistulae in normal and cholesterol-fed rabbits. Pathology 5: 311-324

Stehbens WE (1974a) Changes in the cross-sectional area of the arterial fork. Angiology 25: 561-575

Stehbens WE (1974b) Haemodynamic production of lipid deposition, intimal tears, mural dissection and thrombosis in the blood vessel wall. Proc R Soc Lond (Biol) 185: 357-373

Stehbens WE (1975a) Flow in glass models of arterial bifurcations and berry aneurysms at low Reynolds numbers. Quart J Exp Physiol 60: 181-192

Stehbens WE (1975b) The role of hemodynamics in the pathogenesis of atherosclerosis. Progr Cardiovasc Dis 18: 89-103

Stehbens WE (1975c) Cerebral atherosclerosis. Intimal proliferation and atherosclerosis in the cerebral arteries. Arch Path 99: 582-591

Stehbens WE (1975d) The ultrastructure of the anastomosed vein of experimental arteriovenous fistulae in sheep. Am J Path 76: 377-400

Stehbens WE (1979) Hemodynamics and diseases of systemic blood vessels. In: Stehbens WE (ed) Hemodynamics and the blood vessel wall. Thomas, Springfield, Ill. pp 294-427

Stehbens WE (1981a) The complications of spontaneous and experimental atherosclerosis. In: Garlick D (ed) Festschrift for F.C. Courtice. University of New South Wales, Sydney pp 151-158

Stehbens WE (1981b) Chronic changes in experimental vascular and fusiform aneurysms in rabbits. Arch Path 105: 603-607

Stehbens WE (1981c) Chronic vascular changes in the walls of experimental berry aneurysms of the aortic bifurcation in rabbits. Stroke 12: 643-647

Stehbens WE (1981d) Predilection of experimental arterial aneurysms for dietary-induced lipid deposition. Pathology 13: 735-748

Stehbens WE, Karmody AM (1975) Venous atherosclerosis associated with arteriovenous fistulas for hemodialysis. Arch Surg 110: 175-180

Discussion

WURZINGER: Dr. Stehbens, coming back to your figure showing the arteriovenous fistula, did I get it right that the degree of lipid insudation into the venous wall is most pronounced just opposite the fistula, i.e. where the arterial blood stream vertically hits the venous wall?

STEHBENS: The jet lesion itself does not have all that much lipid deposition in the cholesterol-fed animal.

WURZINGER: I got the impression that there was a more intensive red staining just opposite the fistula.

STEHBENS: I am sorry I did not point out the site of the fistula. In general the lipid deposits were less at the site of the jet lesion, and you do get jet lesions forming very early in the vein in these models.

DEWEY: In the photographs where you had those nice dye traces going down the tubes, you mentioned that the Reynolds number was 55 in the branch vessels, and you also mentioned that there was periodic vortex shedding. I would like to comment that, as far as most arterial flows are concerned, those Reynolds numbers are a little low, and you probably would not see in real arteries the kind of vortex shedding that you saw in those nice smooth pipes. I think it would tend to break down into a more complicated flow, and you would not have a single frequency to the vortex shedding phenomenon.

STEHBENS: It might not be a continuous vortex shedding; however, I do not think at the present time that you can exclude the possibility that they exist. They not only occur at higher Reynolds numbers in the gross form, but even at lower Reynolds numbers there is segmentation of the dye. An important feature is that this change at the forks is augmented by any external disturbances, very much the same as a jet.

A Fluid-Dynamicist's and a Physiologist's Look at Arterial Flow and Arteriosclerosis

A. Naumann and H. Schmid-Schönbein

I. Introduction

The regular localization of early, as well as advanced atherosclerotic lesions in the vicinity of arterial branchings as well as the established role of aging and of arterial hypertension in the progression of this disease has long been recognized (e.g. Aschoff [1]). Consequently, fluid-dynamic considerations have for a long time been included in the pathogenetic theories about the initiation, progression and complications of atheromatosis. Descriptive terms such as "turbulence", "blood stasis", "dead water zones" were used in the early literature when the role of non-laminar flow near branches of the large arteries was discussed.

Later, there was an attempt to replace these phenomenological terms by more quantitative considerations, which were based on a more detailed fluid-dynamic analysis of the flow near bifurcations in vivo or in models of the arteries. These macrorheological concepts resulted in a fluid-dynamic controversy that lasted for more than a decade, in which either *high* wall shear stresses or *low* shear conditions near bifurcations were held responsible for the localization of atheromatous alterations. The concept that high shear stresses might cause atherosclerosis damaging the arterial endothelium was often contradicted by students of the detailed topology of the more advanced lesions (Walløe, this volume). These lesions appear to occur preferentially in vascular areas conventionally associated with the occurance of low shear stresses; however, no convincing mechanisms has been advanced that might be operational in these areas in atherogenesis.

The unresolved controversy between the proponents of the two macro-hemodynamic concepts can most likely be resolved by the knowledge gained in more recent microhemodynamic studies of "disturbed flow" in models of the arteries. Following the pioneering work of Goldsmith [12] and Müller-Mohnssen [22], Liepsch [18] and Karino (communication to this symposium), Baldauf et al. [2], Kinder et al. [15] have identified details of the secondary flow near bifurcations and stenoses. All these authors have shown that *complex*, yet *completely regular* flow patterns with stagnation point flow, local areas of high shear in close vicinity to areas with low shear and recirculation zones occur. Thus, using a more stringent terminology, we are presently experiencing the renaissance of the classical views attributed pathogenetic roles to non-laminar flow and impingement of the flow upon the arterial wall in flow dividers.

Research from which these concepts are derived are supported from grants by Deutsche Forschungsgemeinschaft (SFB 109, C 7) and by Fritz-Thyssen-Foundation (Cologne)

Fluid Dynamics as a Localizing Factor for Atherosclerosis
Editors: G. Schettler et al.

As can be seen in the illustrations to the communications of Liepsch and Karino, arterial bifurcations are characterized by region of accelerated flow velocity and thence abnormally high fluid shear stresses, by flow separation and formation of recirculation zones and by stagnation point flow which occurs in two characteristic points: the vertex of the flow divider and at the reattachment point of the separated stream lines. In the latter, particles carried in the reattaching stream lines are moving forward, whereas particles moving in the closed streamlines of their circulation zone move backward. Particles or molecules carried in the fluid lamellae between those hit towards the wall, these fluid lamellae being a rich source of material to be deposited onto the wall in the immediate neighbourhood of the actual reattachment point. In pulsatile flow, i.e. when the length of the separated flow region and thence the recirculation zone changes during each systole (and may even totally disappear in diastole), the reattachment point is migrating, thereby covering a characteristic area of the arterial wall *downstreams* of the arterial bifurcation.

The revolutionary finding of the microrheological studies of flow in bifurcations are of great practical significance for three reasons:

1) As shown by Rodkievics [31], reattachment points in the natural arterial tree of man coincide with the predelection sites for enhanced permeability to albumin in early and to a lesser extent advanced atheromatous lesions.
2) The structure and biology of endothelial lining, i.e. the shape and orientation of the endothelial cells and their rate of mitosis[1] differ significantly from that found elsewhere in the arterial trees, a fact generally held responsible for the enhanced rate of filtration of macromolecules such as albumin and low density lipoprotein.
3) The fluid-dynamic conditions in the areas of disturbed flow are capable of triggering or amplifying key hemostatic events involving the platelets. As shown by Wurzinger et al. and Blasberg et al. (this volume) high fluid stresses as they occur at the boundary between separated and laminar shear are capable of activating platelets directly or indirectly via lysis of red blood cells, thereby triggering the "release reaction" and the "viscous metamorphosis" and demasking procoagulatory phospholipids. In the recirculation zone, ideal conditions for autocatalytic augmentation of the platelet activation and releases as well as the procoagulatory process exist, whereas the reattachment point provides ideal deposition for platelet deposition, especially if the wall is "adhesive" and if the platelets are "activated".

All three considerations are of substantial interest in the current discussion about the pathogenesis of atheromatosis and its complication since in both the "response to injury"-theory (R. Ross [32]) and the thrombosis theory (Duguid [8]) the platelets, their rheological behaviour and their various biological capacities are pivotal. We wish to propose therefore an alternate - or compromise - theory to the opposing theories of Caro [4] and Fry [10]: we propose that the fluiddynamic, cellular and biochemical events associated with stagnation point flow act as primary injury to the normal arterial vessel wall as well as the perpetual and complicating stimulus to the deve-

1 Recent evidence by S.M. Schwartz and G.K. Hansson (this volume) suggests that some of the endothelial cells in these area may be damaged since they become permeable to markers such as tryptan-blue and immune-globuline-G which are normally not capable of passing intact cellular plasma membranes

loping and complicated lesions. Inherent in this theory is an assumption about the localization of the lesions: we propose that the area of endothelial damage and the deposition of platelets and of platelet derived mediators coincides with the migrating stagnation point, the extent of the lesion therefore being an indication of the domain swept over by the migrating reattachment point as the secondary flows pulsate in systole (see Fig. 1).

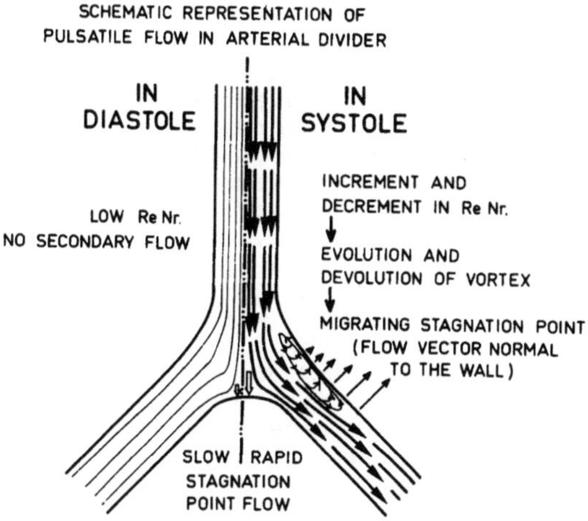

Fig. 1. Schematic representation of the fluid dynamics of pulsatile flow in an arterial bifurcation. *Left half:* during diastole, low flow velocity and thence low Reynold numbers provide pure laminar flow (which can be oscillating in direction); there is a stagnation point flow directed towards the vertex of the flow divider with low normal stresses and shear stresses. In systole, the pulsating flow with an increment and subsequent decrement in Reynolds number leads to an evolution and subsequent devolution of a vortex or recirculation zone. At the reattachment point, there is a flow vector directed against the wall, which migrates peripherally during the acceleration phase and centripetally during the deceleration phase of flow pulse. Depending on flow velocity, there is a more or less pronounced stagnation point flow directed against the divider (*open arrows*)

Before going into the discussion of some pertinent fluid-dynamic (Sect. II) and pathophysiological (Sect. III) aspects of this concept, we wish to dwell shortly on an ontological point we consider essential for the later pathogenetic discussion: it must be taken into consideration that stagnation point flow is a fundamental and inevitable characteristic of fluid motion in branching tube (i.e. arterial) networks - whereas it does not occur in converging (i.e. venous) tube assemblies. There is ample evidence (see [9,34]), that the vertices of bifurcations (or "flow dividers") are characterized by especially high shear stresses and normal pressure due to the skewed velocity profile and the rapid increase in the shear stresses acting in the central flow lines that attach to the inner surface of the daughter vessels. It is not surprising that arterial flow dividers, show *very early* changes at these sites (which may even be present before birth), which have often been classified as pathological (see communication Stehbens). However, while there seems to be no doubt that there are degenerative alterations of the endothelium (see [39] and Buss, this volume) and the subendothelial structures of the so-called arterial cushions (see Stehbens), we wish to propose that these cushions represent nevertheless biological adaptations to an extreme hemodynamic load, which is experienced only by these parts of the vasculature and to which they are exposed from the very development of the vascular system. The cuboidal shape of the endothelial cells found in the vicinity of flow dividers may well be of microrheological advantage since the viscous drag force on each cell (shear stress multiplied by surface area exposed) is considerably smaller than in flat endothelial cells. Furthermore, the more cuboidal shape might allow elastic compliance (and/or viscous damping) of shear stresses varying in direction, magnitude and duration during the systolic peak to arterial blood flow. As stressed by Zarins et al. [39] the intimal pad in the arterial cushions is in fact a highly structured medial zone composed of oriented elastic and smooth muscle layer. The adventitia underlying the flow divider contained many more vasa vasorum than the neighbouring straight vessel segment. The work of Betz [3] suggests that the growth of adventitial vasa vasorum may be secondary to the local increase in endothelial permeability. In summary, then, we wish to propose that the arterial cushions, despite the evidence of cell destruction in them, represents an area adapted to a specialized task of dealing with perpetual vascular damage by stagnation point flow with cell deposition and subsequent wall inhibition with macromolecules and mediators derived from the plasma and the platelets. Conceptionally, this idea allows us to separate *"protected"* from *"unprotected"* arterial regions: we wish to propose that protected areas are "accustomed" to fluid-dynamical stresses and thence structurally and functionally adapted by various alterations of the wall components (Fig. 2). Thus, they differ in their "response to injury" from those

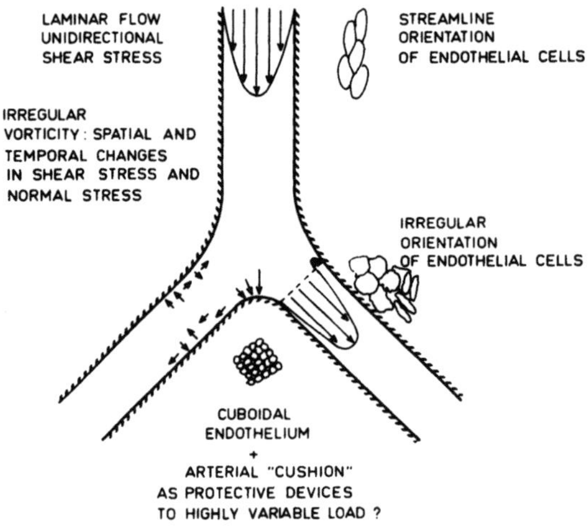

Fig. 2. Schematic representation of the morphogenetic effects of arterial flow on the structure of the intima. Pulsatile flow with phasic acceleration and deceleration leads to periodic oscillations in pressure, velocity, wall shear rate and local and general Reynolds number. In arterial segments with laminar flow, shear stresses acts unidirectionally in diastole and systole, they lead to streamline orientation of endothelial cells, which have a large surface area. The irregular vorticity near the flow dividers results in complicated spacial and temporal changes in the flow forces acting tangentially and normal to the wall. Correspondingly, shape, size and orientation of endothelial cells are irregular. In the area exposed to stagnation points both diastole and systole small cuboidal endothelial cells and subintimal transformations develop during life and may serve as a protective device. Details of the flow are explained in the caption of Fig. 5

areas which are less drastically (or to a later time in life) affected by stress factors, but in which in the absence of adequate functional adaptation a small endothelial damage results in a more pronounced intramural proliferative change. The assumption of local difference in adaptive functions and/or structures could explain an apparent paradox unexplainable by either "low shear" or the "high shear" theory. We propose that the flow dividers are a high shear area but well protected, therefore, while clearly permeable to macromolecules, e.g. to Evans-blue (bound to serum albumin) they could be more fit to cope with intramurally deposited macromolecules and lipids and could thence be relatively spared from the more severe lesions. Put in still other words: the assumption of adaptive changes near the flow dividers can explain why "blueing areas" and fatty streaks regions do not always coincide with the localization of severe and complicated lesions. In the frame of reference laid out by R. Ross ("Response to injury-theory of atherosclerosis") [32] we would like to propose stagnation point flow as the injurious stimulus to the arterial wall, the stimulus strength depending primarily on fluid-dynamic factors, but also secondarily on the composition of the flowing blood and/or of the infiltrate deposited into the subintimal layer. We postulate further that the response to any given stimulus does not only depend on its nature and magnitude, but also upon the reactive potentials of the arterial wall - which in turn reflect the previous history of stimulus exposure and adaptive change elicited by prior exposure to the same stimulus. Thus, the proposed hypothesis could explain advanced steps of atheroma formation as the consequence of both stronger stimulation and of a decompensation of the mural clearing mechanisms in areas damaged fluid-dynamically by sweeping stagnation points in secondary flow in the proximity of arterial branches (Fig. 2).

II. Fluid Mechanics of Secondary Flows (A.N.)

Whatever the biological conception may be about the genesis of arteriosclerosis or of thrombosis, the flow pattern and the forces acting on the fluid flow elements and on the arterial walls appear to play an important role (see [5,6,7,23]). The flow in turn affects the transport, mixing and deposition of elements travelling within the flow. The main questions are, how these elements are carried towards the wall and where the predilected places for this are situated. A velocity to component directed towards the wall in the vicinity of the wall and a certain minimum contact time for the subsequent processes for example the platelet adhesion is therefore indispensible.

a) Laminar or Turbulent Flow? Very often the mixing and the transport are assumed to be caused by "turbulence". However, a certain amount of misunderstanding may exist in the different conception of the turbulence itself, which may be mainly a question of terminology (see also [26]).

The pulsatile laminar flow in tubes with rigid or elastic walls is well known as the Hagen-Poiseuille flow, modified by the pulsation; in laminar flow the mixing takes place by the physical diffusion only. In the general principles of fluid dynamics, the fully developed turbulence is descrived by three dimensional flow disturbances erratically superimposed upon the average velocity, locally and temporally stochastic. Fully developed "turbulence" is different from flow with vortices and eddies: in the branching and curved arteries we have mostly a more or less disturbed flow, characterized for instance by vortices or secondary flows. This type of flow is as far from the laminar as it is from the classical turbulent flow and the established laws of neither flow apply. The appearance of vortices is by no means equivalent to turbulence, and as a risk this apparent turbulence in systole goes back to relaminarisation in the subsequent diastole. It must be stressed that the transition to turbulence requires a certain time, i.e. a certain pathlength of disturbed fluid elements, which is not available in the arteries during the short systole (180-300 ms).

b) The Similarity Conditions . According to the similarity laws all the flow effects are determined by dimensionless parameters; in our case the most important ones are the following: The Reynolds number $Re = \overline{w} \cdot d \cdot \rho/\eta$ represents a measure for the ratio of the inertial to the frictional forces (where $\overline{w}$ = mean flow speed; d = an arbitrary length, e.g. the arterial diameter; ρ = density and η = viscosity of the medium). The Strouhal number $Sr = f\, d/\overline{w}$ is characteristic for the pulsation, where f = frequency of pulsation. The product $Re \cdot Sr = Wo$ is called the Womersley parameter.

$\Delta c/c_0$, $\Delta\rho/\rho_0$: the concentration and the density differences of the suspended elements and the continuous phase. The latter are small for platelets which are practically isopycnic with plasma.

Any pulsation involves further parameters, namely the propagation velocity of the pulse wave and of reflected waves ($\overline{w}^2\rho/E$), which in turn depend on the Young's modulus of blood and wall material resp., and the steepness of the pulse wave flank (d_p/dt).

To establish fluid dynamic similarity with correct boundary conditions geometrical parameters are needed such as the bifurcation angles, the tube length between 2 points of the distance x (x/d, mostly written x/d Re), the curvature radius r/d and others.

c) Flow Past a Strong Stenosis. Flow past a severe stenosis represents a simple and well understood example where the flow behaviour follows from considerations about the flow separation and the formation of the discontinuity layer, the generation of vortices or recirculation and the so-called secondary flows. In order to clarify this, an exaggerated strong obstacle in a tube (Fig. 3) may be considered [25,26,

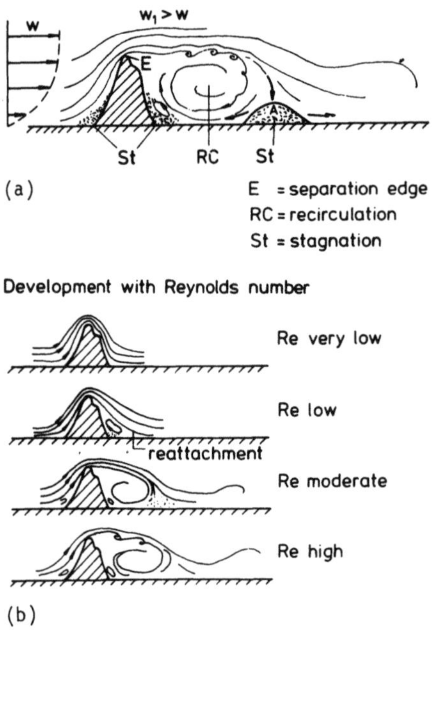

Fig. 3a,b. Schematic representation of the flow patterns produced by a strong arterial stenosis (*St*). Shear stresses are increased in the vicinity of the separation edge (*E*), leading to flow instability downstreams. In the poststenotic area a recirculation zone develops (*RC*). At the reattachment point, a second stagnation point (*A*) may grow, especially if material is deposited by a stagnation point flow

38]. As soon as a critical Reynolds number is reached - its value is lower than that for full turbulence and depends on the form of the stenosis - from the edge E a discontinuity layer is separated; it becomes wavy and rolls up to small vortices (so-called "secondary vortices"). The discontinuity layer represents the limit between the central flow and the recirculation region RC. The normal streamline towards the wall at the end of RC is divided at A, one part of the fluid flows back and carries swimming particles to the stagnation region on the backside of the obstacles. The environment of A is a characteristic stagnation region and can be predilected for an adhesion of material, e.g. sticky platelets.

The recirculation zone is often named deadwater but is by no means dead water, rather we have a complicated flow and the velocities are relatively high. Mostly the sketched flow in the discontinuity layer is laminar, and the small vortices are laminar vortices (i.e. highly structured rather than stochastic). From this situation, the flow can proceed either to full turbulence or back to the complete relaminarisation downstreams, depending on the Reynolds number. In the big arteries neither the first nor the latter is likely to take place.

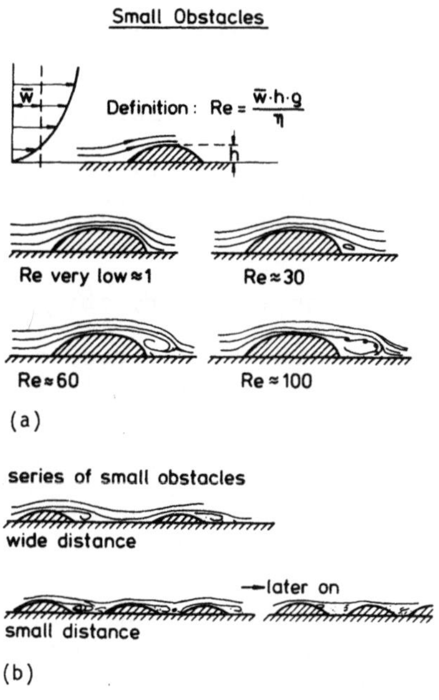

Fig. 4. (a) Development of secondary flow near an arterial stenosis as Reynolds number increase. Flow is laminar at very low Reynolds number, the recirculation zone grows in size and extent as the Reynolds number increases. Therefore, the reattachment point migrates peripherally with increasing flow rate and Reynolds number. (b) Influence of dimension of obstacles and Reynolds number on flow patterns near an irregularly shaped wall. For details see text

Figure 4 shows the development of the flow pattern with increasing Reynolds numbers. At very low ReN - at negligible inertial forces - no separation occurs. At increasing but yet low ReN a small recirculation is observed; then the separation forms the great wake, the boundary layer of which rolls up finally into the small vortices as described before. The corresponding Reynolds numbers are determined by the shape of the stenosis, specially by the detaching ridge.

The forces which are exerted by the flow on the wall in the environment of the stagnation region are qualitatively sketched in Fig. 5

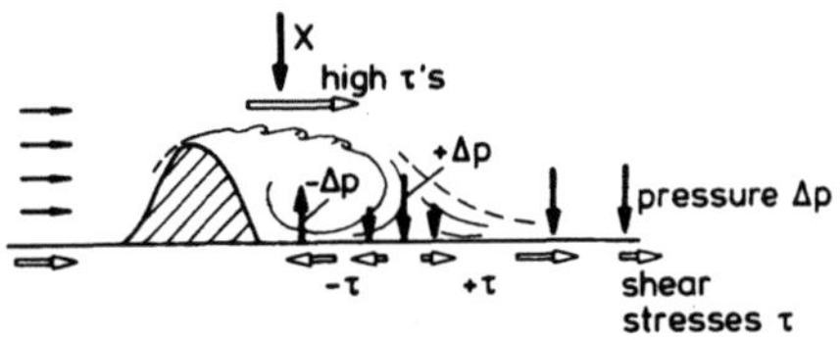

additional forces during systole

(a)

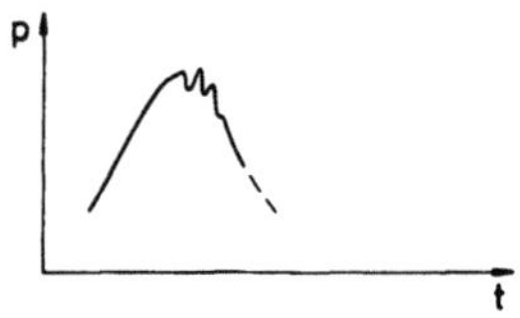

Pressure course against time at X

(b)

Fig. 5a,b. Generation of additional forces in flow past the stenosis in high Reynolds number. The pressure decreases in the recirculation zone and increases behind it. Therefore, there is an additional pressure normal to the wall at the reattachment point and a diminished pressure on the wall in the area of the vortex. These pressure variations oscillate with the flow pulse, pressure pulse and volume pulse generated by the heart *(closed arrows)*. The wall shear stresses *(open arrows)* act in different directions upstreams and downstreams of the reattachment point. Therefore, the wall near the recirculation zone is subject to wall shear stresses that change in magnitude, duration, and direction with each heartbeat

(not in scale!). Near the wall the total pulsatile pressure is given as the sum of the static pressure and additionally of the dynamic pressure, caused by the recovery of the cinetic and the pulsation energy. Certainly, this additional pressure is low compared with the static pressure; but as a long time effect, the increased transmural pressure can lead to a deformation of the wall and in consequence to an aneurysm. We feel that the shear stresses are more effective because they have opposite directions upstream and downstream, so they can become able to stretch or even to disrupt the endothelium (see also [4,10]).

The discontinuity starting from the separation edge is a very thin layer, in which a high velocity gradient normal to the main flow direction exists, this implies high shear stresses, which are able to damage the red blood cells (hemolysis) as it is known from artificial heart valves and artificial organ parts. At the same place the platelets can be activated; the activated cells are transported within the flow and can be deposited anywhere far downstream from their activation place [23,24,25]. It may be mentioned that such high shear stresses cannot occur at the normal arterial wall but only near and at severe stenoses, especially when an arterial stenosis leads to an appreciable pressure gradient.

The flow patterns described so far represent a momentary picture. In the arterial system, the size of the recirculation region pulsates and so does the site of the stagnation point in the reattachment. Each pulse wave corresponds to a pressure impulse and produces a more or less jerky high acceleration, which is considerably influenced by the steepness of the wave flank as mentioned before. This is of particular interest since the pulse pressure increases physiologically with age. Thus, there results an unsteady increase of the momentary forces with steep peaks of shear stress; and vortices are generated already at even lower mean Reynolds numbers. Both these phenomena are equivalent to an effective enlargement of the shear stresses, often accompanied by the increase of the flank steepness with the age dependent hardening of the arterial vessels.

In cases when the stenosis is limited to one side of the artery, the flow past the obstacle has a lateral component and assumes a three-dimensional character, the stagnation line has a limited extension and regional, localized secondary flow occurs.

Unfortunately, there is a limited knowledge only about the influence of the pulsation and of the wave form on the size of the recirculation region as well as on the amount of the effective shear stresses.

As shown recently [11], the presence of an oscillatory component in flow velocity alters the pattern of steady flow around a nonsymmetric constriction, which is characterized by a pair of standing vortices behind a stationary vortex (which resembles the shape of a horseshoe) in front of the obstacle. In pulsatile flow, nonsymmetric vortex rings are generated periodically behind the obstacle and shed downstream in succession, and periodical changes also were observed in the position and width of the resulting vortex.

d) The Flow Past Wall Irregularities. The preceding remarks had to do with a strong stenosis. In principle, the analogous flow behaviour occurs behind smaller stenosis, in curved arteries, in ramifications, at wall deformations as well as in artificial bloodfilled organs. A small stenosis, representing a slight irregularity of the arterial wall (Fig. 4b) produces a separation and a recirculation already at Reynolds numbers as low as about 10. At the reattachment point, normal speed components occur; the "secondary vortices" begin at about $Re \sim 60$. The dimensionless length parameter in the Reynolds number is here the height h of the stenosis. The Reynolds numbers, at which the different stages begin, are thus depending on the geometry of the obstacle, e.g. on the height/length ratio and the course of the curvature of the contour. Furthermore the figure shows the flow past a series of successive obstacles with a wide and a small distance; after a long delay a gradual fill up with foreign material is the consequence at a narrow position. Biological processes occurring at "rough" arteries with complicated lesions are likely to be affected by the repetition of the secondary flows and reattachment points.

e) Bends, Bifurcations and other Configurations. The flow behaviour in curved arteries (bends) is also determined by the occurrence of the "secondary flows" (after Prandtl [30]). The centre stream which has a higher speed than near the wall, does not follow the curved central line of the pipe because of its higher inertia (Fig. 6). Thus, a pressure difference between the outside and the inside of the bend results, an additional flow, a spiral movement and a flow separation takes place. According to the momentum law, a high pressure loss results from the change of the velocity distribution. This situation is likely to occur at the aortic arch.

The course of the stream lines in bifurcations is somewhat more complicated because of the influence of the central core and its shape. Recirculations and stagnations cause increased normal and tangential tensions in quite the same manner as behind stenosis. Here, too, the evident effect is the secondary flow which is superimposed to the longitudinal stream lines. It leads to helical stream lines in the branches and enhances the tendency for a strong separation. In the lower drawing of Fig. 5, the courses of two streamlines a and b outside of the centerline of the tube and the velocity distribution are qualitatively depicted. The helical flow has been first studied by H. Zeller and his coworkers [35,40,41] about 12 years ago. Much work has been done in the last years, here it may be related as examples to the publications of H. Müller-Mohnssen [20,21], D. Liepsch [16,17], and Karino and H.L. Goldsmith [14]; the latter observed the flow in bifurcations at different angles and the oscillation of the recirulation regions.

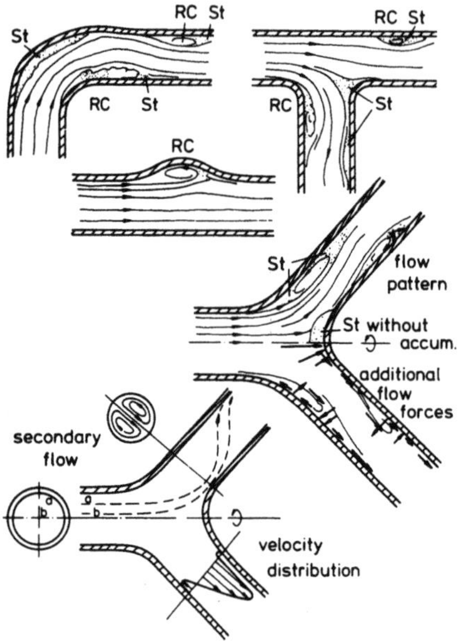

Fig. 6. Schematic representation of secondary flows in curved, branching and locally dilated arteries: all geometric features shown give rise to secondary flows. The local details of pressure, shear stress, velocity distribution and stagnation points (with or without accumulation) are sketched (for explanation see caption of Fig. 5)

III. Physiology and the Pathophysiology of Pulsatile Arterial Flow

According to the presently popular "response to injury theory" of atherogenesis (R. Ross) fluid-dynamic forces (for example stagnation point flow) exert an injurious effect upon the arterial endothelium, which in turn leads to primary reactions such as

1) filtration of plasma lipids and/or lipoproteins into the arterial wall, and
2) platelet deposition and other thrombotic reactions as well as the liberation of platelet derived mitogenic substances.

These processes are assumed to give rise to secondary changes, such as phagocytosis of plasma lipids by transformed smooth muscle cells, cell migration and proliferation. The impact of a localized stagnation point flow on the permeability of endothelium was analyzed by Patel et al. [28] who modelled such fluid-dynamic forces with a saline yet directed towards the arterial wall. Their results confirm that localized shear stresses in the order of magnitude of 40 Pa (400 dyn/cm^2) lead to an area of elevated permeability the shape of which reflected the distribution of shear stresses near a stagnation point. These authors concluded that the damaged region corresponds to the field where the incident shear stresses are too large to be sustained by the endothelium. The assumption that secondary flows and stagnation point flow (rather than high or low laminar shear) may exert injurious effects, bears the possibility that changes in the pulsatility of pressure and flow in arteries, such as they occur physiologically with aging, may represent important atherogenetic determinants which have not received wide attention in the past.

If one accepts that the stagnation point flow (where flow vectors are directed towards the arterial wall) and the recirculation zones (where frequent collisions occur between blood born molecules and cells) are of pathogenetic significance in atherogenesis, it follows that such deviations from simple laminar flow exert such action not only in fully developed, but already during the development and still during the regression of secondary flow. Since secondary motions in pulsatile arterial flow depend on a great number of factors [29], the function of the left ventricle as a pump, composition and compliance of the arterial wall, the geometry of arteries and of the bifurcations, the peripheral resistance, and the viscoelasticity of the blood as a non-Newtonian fluid all exert decisive influences since they all contribute to the temporal and spatial variations in flow and pressure, wall shear rate and wall shear stress and general and local Reynold's number (as depicted schematically in Figs. 1,2,7). It is of great interest that all parameters listed change in the circulatory systems of man during the normal process of aging.

To the best of our knowledge, the in vivo analysis of the arterial fluid-dynamics has usually been performed on the basis of measurements in young experimental animals, presumably with highly distensible arteries which physiologically damp out the pulsation in flow and pressure. This reservation also holds for Nerem's [26a] pioneering studies on arterial flow in horses; in addition, the hot-film anemometry unfortunately is not a suitable method to detail the nature of the flow instabilities he detected in the large arteries of conscious horses.

Aging is - in all animals and in man - associated with the arterial hardening (v.i.). Atherogenesis in turn is a slow process that occurs primarily in the *aging* human arterial tree. Owing to the fundamental laws of fluid-dynamics, it is unevitable that the pulsatile arterial flow in the aging human vasculature physiologically undergoes very pronounced changes, for example a tripling of pulse pressure difference between peak systolic and end-diastolic arterial pressure. This results from the fact that in man and probably in most animals the size of the arterial vessel increases with age while at the same time elastic wall components are replaced by the much stiffer collagen. As the "Windkessel-Funktion" is last during aging not only to the well known augmentation of the pulse pressure occurs but also to changes in the propagation of both pressure and flow pulse into more peripheral parts of the arterial tree. The literature on this topic, unfortunately, is not too extensive, the need or more detailed analysis of these physiological changes during the normal process of aging has been stressed in a recent review of O'Rourke. In quoting O'Rourke verbatim we would like to endorse his call for more research because "arterial 'degeneration' with age is so common in man as to be a practically normal phenomenon. Already more than a century ago it was stated that "only in the case of young children .... is the elasticity of arteries .... as perfectly adapted .... as in the case of lower animals". While admittedly Nerem's measurements of "transitional flow" in healthy horses are representative for healthy animals of similar size (and thence to human adults) they do not allow any predictions about the fluid dynamics of pulsatile flow in the aging vasculature.

In the absence of detailed information about the age related changes in the flow and pressure pulses in the peripheral parts of the arterial tree, we have to resort to speculation which, however, is well supported by measurements of the arterial elasticity, speed of pulse wave propagation and geometry (for a review see O'Rourke 1982). Cardiologically speaking, these changes are in principle "compensated" by an adaptation of the cardiac muscle to the resulting increase in "arterial impedence".

Inherently, the cardiac muscle is capable of adapting to the resulting increase in "afterload"; the heart is forced, however, to procure a larger part of its contractile energy to overcome the inertia of the blood. In the aged, not only the mass equivalent to stroke volume, but also equivalent to the volume residing in the aorta and the central arteries has to be accelerated in systole. Up to 26% (O'Rourke [27]) of the cardiac work must be expended into kinetic energy during systolic acceleration, which results in high inertial effects - which increase the ReN - in the more peripheral arteries. Inertial effects in arterial flow are mostly "lost" from the pressure energy that drives the blood through the peripheral resistance vessels, at least part of this energy is transmitted onto the walls of the aging vasculature and must act there by increasing the locally effective pressure pulse and flow pulse. The resulting alteration in the local shear stresses, normal pressure and flow components towards the wall near bifurcations and stenoses await elucidation. It is justified, we believe, to postulate that the augmentation of flow and pressure pulsatility must increase the putative fluid-dynamic damage to the endothelium in the aging vasculature. In addition, we believe, that it is justified to speculate further that any rise in peripheral resistance or in stroke volume (as it is known to occur in patients with essential hypertension (see textbooks of cardiology)) must further enhance the pulsatility of arterial flow as it either increases or decreases diastolic flow. It follows from these considerations that a combination of arterial stiffening and elevated peripheral resistance must lead to especially pronounced augmentation of secondary flow at arterial bifurcations and thence lead to a more effective injury by normal and tangential forces acting on the endothelium (v.s.).

There are local differences in pulsatile flow that must be considered. Atherosclerosis progresses with age - but unfortunately the rate of progression seems to be higher in the coronary and carotid arteries than in other arteries. Assuming that the flow is indeed a causative factor in the rate of progression, it is important to note that the arterial fluid-dynamic in the coronary and carotid artery are significantly different from that in the arteries supplying other organs: they are both characterized by especially pronounced pulsatility and thence in their susceptability to secondary flows. In the coronary artery, the systolic interruption of flow not only produces exceptionally high flow rates in diastole - but these are further enhanced in cardiac hypertrophy - which often occurs in essential hypertension. Furthermore, if the heart is under sympathetic drive, it not only further increases blood flow - but also experiences an overproportional increase in diastolic flow. In diastole, therefore, not only an overproportional flow ($\dot{V}$), but even more so an overproportional increase in $d\dot{V}/dt$ or $d\dot{V}/dt^2$ must be expected, i.e. a very rapid increase in early systole ReN.

In all other arteries and especially in the carotid arteries, the pulsatility is reversed: here flow is accelerated primarily during systole. It has been shown by a number of techniques that in the internal carotid arteries diastolic flow decreases even in normal aging subjects - but even more so in patients with signs of cerebrovascular insufficiency [19]. This also results in a more pronounced flow pulsatility and thence higher systolic peak ReN. Since cerebrovascular resistance, i.e. the input impedance to the internal carotid artery increases with age - simply because cerebral blood flow is reduced with increasing age even in normal subjects, more and more complicated flow patterns are likely to occur at the level of the carotid bifurcation as a subject ages. All these changes in the arterial pulsation are likely to be more pronounced if complicated by general "physioclerosis" and by elevated peripheral resistance. Both factors enhance peak systolic flow acceleration in the common carotid arteries; thence it can be postulated that the putative flow damage to the arterial endothelium must be more pronounced in aging patients which are carriers of so-called risk factors, such as hypertension, diabetes and frequently in smokers.

Last no least, it is of interest to emphasize that for any given ReN and "distensibility", the viscoelastic properties of the blood appear to be important determinants of the absolute shear stresses that occur in the areas of secondary flow and near the stagnation point flow. The viscoelastic properties of the blood critically depend on the hematocrit and the tendency to red cell aggregation, i.e. the formation of elastic structures with the blood based on reversible interactions between adjacent red cells in so-called rouleaux (for reviews see [13,37]). As shown by Liepsch, the resulting change in the velocity profile in the sense of axial (blunting) increases the wall shear stresses. There is ample evidence that the viscoelastic properties of the blood are also more pronounced in carriers of risk factors for atherosclerosis and its complication [33].

As was shown above, the formation of secondary flow is greatly facilitated by macroscopic "roughness" of arteries, i.e. by the consequences of existing atherosclerotic changes. It needs no emphasis that the combination of such a local change with the general fluid-dynamic alterations described above is likely to potentiate the effect of either abnormality alone.
Thence, if one accepts that atherosclerotic lesions become more elevated with increasing age and are more irregularly distributed over the arterial tree, one must accept that the fluid-dynamic effects of advanced lesions are likely to become more effective with age. Even if one rejects the very idea that the early alterations (such as of increased endothelial permeability and fatty streaks and fibrous plaques) are caused by fluid-dynamic factors, late and progressive wall alterations must be accepted as arterial sites, which gives rise to localized flow abnormalities that favour thrombotic deposition processes. It follows from general fluid-dynamics that surface irregularities and especially all lesions which strongly reduce the arterial diameter must lead to the described flow disturbances which, in turn, must be more pronounced under all conditions in which the pulsatility of arterial flow and pressure is enhanced.

Summary

This paper presents a concept that assumes "stagnation point flow" in pulsatile arterial flow acts as the injuries stimulus on arterial endothelium as postulated by R. Ross.

Inherent in the response to injury of atherosclerosis theory is the assumption that fluid-dynamic forces might act as mechanical injury to the endothelium. We would like to propose that the occurrance of stagnation point flow coincide with the preferential occurrance of early lesions near bifurcations of the arterial tree. We assume further that the arteries develop protective mechanisms in the so-called arterial cushions which develop in the region of the flow divider in which pronounced stagnation point flow occurs during the entire life from the very onset of arterial flow in the developing embryo (Fig. 7).

After an introductory overview of the fluid-dynamics of secondary flows and reattachment points in pulsatile flow through stenoses, bends and bifurcations, circulatory factors which affect secondary flows are discussed in their ability to act as an injurious stimulus to the endothelial cells. In this context, the instabilities of flow and pressure in pulsating arterial blood motion may acquire a pivotal pathoge-

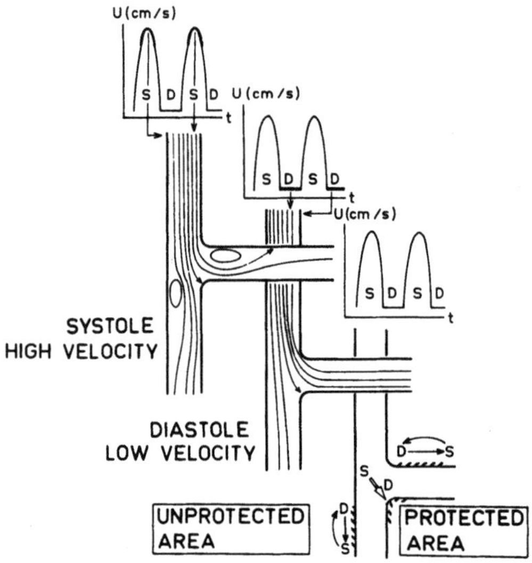

Fig. 7. Summary of the fluid dynamic and biological concepts of the present communication in which flow towards the wall (which occurs in flow dividers and at reattachment points) is considered as the damaging factor for the endothelium. The flow divider is protected from progression of atherosclerosis because of adaptive mechanisms which develop early in life. *D* = diastole; *S* = systole, for details see text

netic role. The consequence of the well established age-dependent augmentation of arterial pressure pulsatility are stressed and related to naturally occuring changes in the arterial wall (mostly the media) of the hardening "Windkessel" vessels (aorta and large arteries). The resulting increase in arterial systolic pressure and pulse pressure and a more rapid increase in pressure (dP/dt) are discussed. The role of adaptive changes in the heart to the increased input impedance of the aged arterial tree (review see O'Rourke) and on the *flow* pulsatility in the large arteries are stressed, as well as the role of enhanced pulsatility.

References

1. Aschoff L (1912) Thrombose und Sandbankbildung. Beitr path Anat 52: 205-212
2. Baldauf W, Wurzinger LJ, Kinder J (1978) The role of stagnation point flow in the formation of platelet thrombi on glass surfaces in tubes with various geometry. Path Res Pract 163: 9-33
3. Betz E, Schlote W (1979) Responses of vessel walls to chronically applied electrical stimuli. Basic Res Cardiol 74: 10-20
4. Caro CG, Fitz-Gerald JM, Schroter RC (1971) Atheroma and arterial wall shear - observation, correlation and proposal of a shear dependent mass transfer mechanism for atherogenesis. Proc R Soc Lond B 177: 109-159
5. Caro CG, Pedley TJ, Schroter RC, Seed WA (eds) (1978) The mechanics of the circulation. Oxford University Press, New York Toronto
6. Doerr W (1963) Perfusionstheorie der Arteriosklerose. Thieme, Stuttgart
7. Dropmann K (1963) Über die Prädilektionsstellen der Atherosklerose und die hämodynamischen Verhältnisse in Krümmerstrecken. Z Kreislauff 52: 171-183
8. Duguid JB (1946) Thrombosis as a factor in the pathogenesis of coronary atherosclerosis. J Path Bact 58: 207
9. Friedman MH (1974) Shear profiles and diffusion in idealized flow through a y-branch. In: Nerem RM (ed) Fluid dynamic aspects of arterial disease. Ohio State University, pp 1-8
10. Fry DL (1968) Acute vascular endothelial changes associated with increased blood velocity gradients. Circ Res 11
11. Fukushima T, Azuma T (1982) Patterns of pulsatile flow in arterial models with stenosis. Clin Hemorheology 2: 31-41

12. Goldsmith HL, Yu SSK, Marlow J (1975) Fluid mechanical stress and the platelet. Thromb Diathes Haemorrh 34: 32-41
13. Kaibara M, Fukada E (1982) Transient viscoelastic behavior of blood. Clin Hemorheology 2: 7-11
14. Karino T, Kwong HM, Goldsmith HL (1979) Particle flow behaviour in models of branching vessels. Biorheology 16: 231-248
15. Kinder J, Kratzer M (1975) Geschwindigkeitsmessung im Inneren komplizierter Strömungen mit einem Lichtschnittverfahren. Biomed Technik 20: 11-12
16. Liepsch D (1974) Untersuchung der Strömungsverhältnisse in Verzweigungen von Rohren kleiner Durchmesser (Coronararterien) bei Stromtrennung. Dissertation TU München
17. Liepsch D (1978) Sichtbarmachung der Strömungsvorgänge in Arterienmodellen bei stationärer und pulsierender Strömung. Biomed Techn 23: paper 139
18. Liepsch D, Moravec St (1979) Qualitative und quantitative Strömungsuntersuchungen an einem menschlichen Nierenarterienmodell. Z Biomed Technik 24: 184-191
19. Mol JM, Rijken WJ (1974) Doppler haematotachographic investigation in cerebral circulation disturbances. In: Reneman RS (ed) Cardiovascular applications of ultrasound. Elsevier Publishing Comp., New York, pp 305-313
20. Müller-Mohnssen H (1957) Über hydrodynamische Ursachen der Arteriosklerose und Thrombenlokalisation in den Coronararterien. Beitr path Anat 117: 283
21. Müller-Mohnssen H (1958) Die Strömungsverhältnisse in den Coronararterien und ihre Bedeutung für die Manifestierung der Coronarsklerose. In: Probleme der Coronardurchblutung. Springer, Göttingen , S 179-196
22. Müller-Mohnssen H (1971) Pathogenese der Koronarsklerose und Strömungsmechanik. Münch Med Wschr 113: 604-616
23. Naumann A (1969) Strömungsfragen der Medizin. Arb Gem Forschg Nordrh Westf 203
24. Naumann A (1975) Strömung in natürlichen und künstlichen Organen und Gefäßen. Klin Wschr 13: 1007-1019
25. Naumann A (1981) Strömungsfragen der Biotechnik. 17. WEH-Seminar "Naturwissenschaftliche und Medizinische Aspekte der Biomaterialien", Bad Honnef
26. Nerem RM (1977) Hot-film measurements of arterial blood flow and observations of flow disturbances. In: Hwang NHC, Normann NA (eds) Cardiovascular flow dynamics and measurements. University Park Press, Baltimore London Tokyo, pp 191-215
27. O'Rourke MF (1982) Vascular impedance in studies of arterial and cardiac function. Physiol Rev 62: 570-623
28. Patel DJ, Vaishnav RN, Atabek HB (1979) Local mechanical properties of the vascular intima and adjacent flow fields. In: Hwang NHC, Gross DR, Patel DJ (eds) Quantitative cardiovascular studies. Clinical and research applications of engineering principles. University Park Press, Baltimore, pp 215-231
29. Pedley TJ, Schroter RC, Sudlow MF (1971) Flow and pressure drop in systems of repeatedly branching tube. J Fluid Mech 46: part 2, 365-386
30. Prandtl L (1969) Führer durch die Strömungslehre, 7. Aufl. Vieweg, Braun - schweig
31. Rodkiewicz CM (1978) Brief discussion: Possible separation and stagnation regions of arterial tree. In: Nerem RM, Cornhill JF (eds) The role of fluid mechanics in atherogenesis. Ohio State University, pp 14-1 - 14-4
32. Ross R (1981) Smooth muscle cells and atherosclerosis. In: Moore S (ed) Vascular injury and atherosclerosis. Marcel Dekker, New York Basel, pp 53-77
33. Schmid-Schönbein H, Volger E, Teitel P, Kiesewetter H, Dauer U, Heilmann L (1982) New hemorheological techniques for the routine laboratory. Clin Hemorheology 2: 93-105
34. Smith KA, Colton CK, Freedman RW (1974) Shear stress measurements at bifurcations. In: Nerem RM (ed) Fluid dynamic aspects of arterial disease. Ohio State University, pp 12-15
35. Talukder N (1974) Untersuchung über die Strömung in arteriellen Verzweigungen. Dr.-Ing. Dissertation, RWTH Aachen
36. Talukder N, Nerem RM (1978) Flow characteristics in vascular models. Int Conf Mech in Medicine and Biology, Vol VII, p 281

37. Thurston GB (1981) Significance and methods of measurement of viscoelastic behavior of blood. In: Gross DR, Hwang NHC (eds) The rheology of blood, blood vessels and associated tissues. Sijthoff & Noordhoff, Alphen aan den Rijn, pp 236-256
38. Tomm D (1978) Strömung und Geräusch in verengten und verzweigten Gefäßen. Dr.-Ing. Dissertation RWTH Aachen
39. Zarins ChK, Taylor KE, Lundell MI, Glagov S (1978) Aortic ostial morphology and the localization of atherosclerotic lesions: Preliminary observations. In: Nerem RM, Cornhill JF (eds) The role of fluid mechanics in atherogenesis. Ohio State University pp 5.1-4
40. Zeller H, Talukder N, Lorenz J (1970) Model studies of pulsating flow in arterial branches and wave propagation in blood vessels. AGARD Conf Proc 65, paper 15
41. Zeller H, Talukder N, Lorenz J (1970) Versuche zur pulsierenden Strömung durch arterielle Verzweigungen. Abh Aerodyn Inst Aachen 20: 44-48

Discussion

BORN: To the best of my knowledge there is really no evidence that catecholamines are ever, even under maximal stress, at concentrations high enough in arterial plasma to affect the arterial wall. The concentrations are exceedingly low, even after a very severe coronary haemorrhage.

SCHMID-SCHÖNBEIN: I am aware of several studies where even in young subjects physical exercise or physiological stress is sufficient to activate, for example, platelets. If I am correct, and we are here in one of the meccas of this field, I think that the peak concentrations of total catecholamines approach 10^{-9} rather than 10^{-11} molar, and this is certainly sufficient to activate the platelets in our system. Also I think there is little doubt but that the diabetic produces more catecholamine output for a given amount of physical exercise.

Stehbens: I would like to ask Prof. Schmid-Schönbein and also Prof. Naumann one question. In the microcirculation, the central or axial zone, consisting primarily of red cells, is separated from the wall by the marginal or plasma zone, which apparently widens with an increase in the velocity. Is there any information as to what actually happens in larger arteries? Is there a very wide plasma zone in the sort of arteries we are talking about?

SCHMID-SCHÖNBEIN: We will hear more about this in the afternoon; however, methodological reasons have so far precluded the study of this, and it is an open question.

SCHETTLER: You mentioned that cigarette smoking might influence the fluid dynamics and this takes us back to catecholamines. Dr. Schmid, Dr. Dietz and Dr. Schömig, from our department and Dr. Kübler's department, did some investigations on the catecholamine levels after heavy smoking, and they did not find any differences. It sometimes has been described that there was an increase of catecholamines after smoking, but we are unable to find that, and it might be that it is a matter of methods. Could it be that prostaglandin synthesis is affected by smoking and other methods?

SCHMID-SCHÖNBEIN: I think I confused matters by having one diagram which showed sympathetic activity and another one which showed increased catecholamine levels in the circulating blood. I think there is little doubt but that smoking increases sympathetic drive and that we have what has been called in German "frustrane cardiovasculäre Aktivität" which the best translation is "futile cardiovascular activity". In other words, we have an increase in heart rate at the same time there is an increase in peripheral resistance, which of course leads to an increase in mean arterial pressure which partly of course is damped by the regulatory system. But there is very little doubt that this is a situation where, from simple theoretical reasoning, we must expect a more pronounced pulsatility of the arterial flow. If this happens very often simultaneously to the physiological aging, this I believe could supply a basis for a fluid dynamic interpretation of the risk factors which were established by people who are not really interested in mechanisms.

The Case for Fluid Dynamics as a Localizing Factor in Atherogenesis

R. M. Nerem and M. J. Levesque

A. Introduction

The possible role of fluid dynamics in the localization of atherosclerotic lesions has provided a primary motivation for the resurgence of interest in the study of blood flow through the major mammalian arteries which has taken place over the past 15 years. Such flows are pulsatile and pass through vessels which are complex both in geometry and in their elastic nature. It is the former which is of particular interest with regard to atherosclerosis. It now appears that it is this geometric complexity, as represented by vessel branching and the general tortuosity of the vasculature, which may be all important, not only in determining the detailed characteristics of arterial flows, but also the spatial nature of any fluid dynamic involvement in the disease process.

Of course severe atherosclerosis produces changes in the geometry of the arteries involved, and in so doing alters and interferes with blood flow. This is particularly important for the heart and the brain, with the result being myocardial or cerebral ischemia or even a myocardial infarction or stroke. However, even in the early stages of the disease process, i.e. during atherogenesis, it is believed that there is an important relationship between the disease and the characteristics of the blood flowing through the arteries. In these early stages of the disease, the vascular geometry is unaltered from its normal state, although this normal geometry may vary considerably from individual to individual. If there is a role of arterial dynamics in the early development of atherosclerosis, it thus is the fluid dynamics of the normal cardiovascular system which is of interest. Furthermore, considering the characteristic time of this disease, the role of fluid dynamics is quite possibly an extremely subtle one. This suggests that it is the rather detailed flow characteristics which may be important, and it is the evidence for such a subtle involvement, the mechanisms that may be participatory, and the detailed flow characteristics themselves that are discussed here and used as a basis for stating the case for fluid dynamics as a factor in lesion localization.

B. Evidence for a Fluid Dynamic Involvement

The evidence for the involvement of fluid dynamics in the atherosclerotic process centers primarily on the pattern of the disease [1-3]. Not only is it the aorta as well as the iliac, femoral, coronary, and cerebral arteries which are most commonly affected, but to be more specific, it appears that it is often regions of arterial branching and sharp curvature which have the greatest predilection for the development of atherosclerosis. These are also regions where the flow will assume unusual characteristics or at least deviate from what otherwise might be considered a well-behaved arterial flow. It thus is the indictment provided by this indirect evidence, particularly as

Fluid Dynamics as a Localizing Factor for Atherosclerosis
Editors: G. Schettler et al.

it relates to the bifurcations and geometrical contortions of the arterial vasculature, which motivates much of the present interest in arterial fluid dynamics.

Unfortunately, in terms of understanding the hemodynamics as a factor in the atherosclerotic process, it is not enough to identify the general regions of branching and sharp curvature. This is because such general regions may include locations of flow separation, localized positions of accentuated secondary motion, and areas where the wall shear stress will be high and others where it will be low. Although there is limited information available on the specifics of the pattern of the disease in man and animals, the available results seem to divide into two groups. In human fetuses, neonates and infants [4] and in experimental diet-induced animal studies [5], the pattern of the disease appears to favor the occurence of lesions in regions which, as we will see in a later section, are believed to be high shear. However, in adult human disease [3,6,7] and in the White Carneau pigeon [8] where spontaneous atherosclerosis occurs, the disease pattern appears to favor low shear regions. This difference in spatial location between lesions raises important questions, not only with regard to the possible importance of fluid dynamics in atherogenesis, but also in terms of differences in the development of experimental as opposed to spontaneous disease.

In addition to the pattern of the disease, there is another piece of evidence with potential fluid dynamic implications. This is that hypertension is a risk factor in coronary artery disease. A basic question, the answer to which has considerable implication for fluid dynamics, is what is the role of blood pressure from a mechanistic point of view? At one time it was thought that blood-arterial wall lipid transport was due to a pressure driven, bulk flow process. However, data now available tends to refute any major importance of such a pressure effect [9]. Pressure still could have an effect on the endothelium or on wall metabolism. Furthermore, if it is an elevated pulse pressure that is important in hypertension being a risk factor, then associated with such an increased pulsed pressure, there would be an altered velocity waveform, an increased peak velocity, and an increased level of wall shear stress [10]. However, if it is mean blood pressure that is important in terms of a hypertension effect, then it may be due to its effect on arterial geometry. With regard to the latter, it has been reported that gross elevation of aortic pressure distorts the geometry of the aortic tree to the point where angles of branching are markedly changed [11]. This could have the effect of changing a relatively streamlined flow system into a considerably more tortuous one which would tend to accentuate the type of complicated fluid dynamic phenomena already present.

There are also data showing an accelerated occurence of atherosclerosis in subjects with a human coronary geometry deviating from that normally found [12]. Furthermore, aortic coarction, which represents a major change in vessel geometry, results in a significant alteration in the uptake of Evans Blue Dye [13], and our own recent studies have demonstrated a similar effect on endothelial geometry and orientation.

In considering the various factors involved in atherogenesis, the one that clearly and most directly would be influenced by arterial geometry would be the detailed fluid dynamic characteristics. This suggests that vascular geometry and the way it influences the local detailed flow properties may be a primary determinant of a hemodynamic effect on atherosclerosis, and this will be explored further in the next few sections.

C. Fluid Dynamic Influences on the Endothelium

Studies of mechanisms by which fluid dynamic-related events could enter into the atherogenic process have focused on the endothelium and, in particular, on transendothelial transport. Fry's early results [14] suggested that the hemodynamically imposed wall shear stress influences the rate of transport of macromolecules between blood and the arterial wall. However, the results of Caro and Nerem [15] using ^{14}C-4-cholesterol have shown that it is not an effect of wall shear on the diffusion boundary layer transport, but one considerably more subtle in the terms of an effect of shear stress on the properties of the arterial wall itself. Similar steady flow results have been obtained using ^{131}I-albumin, and data from both radioactively labelled albumin and cholesterol experiments indicate that it is not only the mean component of arterial pressure and flow, but also the oscillatory nature of the phenomena which has an important influence on the blood-arterial wall transport of macromolecules [16,17]. These results suggest that, at a wall shear stress level greater than 50 dynes/cm^2, there is a much stronger dependence of the transport rate on shear stress [10], and this has been confirmed by the more recent results of Thibault et al. [18].

There also has been considerable interest in the effect of fluid dynamic forces on the endothelial cells themselves. One question being asked is whether or not physiologic wall shear stresses are large enough to erode endothelial cells from the wall, i.e. to cause denudation. Fry's results [14] indicate that an erosion stress level of approximately 1000 dynes/cm^2 would be required to cause denudation. However, a recent study by Vaishnav et al. [19] indicates that there are important effects associated with temperature and duration of exposure, and these results suggest that for short exposure times, the critical and/or erosion stress levels may be considerably higher. This is also suggested by the results of Languille [20]; and these higher stress levels would appear to be non-physiologic.

Our own effort has focused on the influence of fluid dynamic forces on endothelial cell geometry and orientation. If the functioning of a cell is influenced by its shape, then this could be a means whereby a fluid dynamic effect on the atherogenic process becomes manifest. Whether this would support a high or low shear interpretation of the pattern of the disease would of course depend on the exact nature of any relationship between cell function and cell shape. Using the vascular casting technique, we have studied the influence of pressure on endothelial cell shape in rabbit aortae. These data indicate that there is a strong pressure effect, with endothelial cells becoming rounder and less elongated with increasing pressure [21]. Also using the rabbit, we have investigated endothelial cell patterns in the region of intercostal ostia. These results have been shown to be very much suggestive of the complicated flow patterns that one might expect to be present [22]. Recently in vitro studies of the endothelial cellular dynamics have been initiated using cultured populations of vascular cells. Dewey et al. [23] have reported the use of a rotating cone-plate viscometer to study cultured bovine aortic endothelial cells under conditions of a uniform fluid shear stress. In our own laboratory we are using a parallel plate, channel flow device for similar studies. Our initial studies also have involved the use of cultured bovine aortic endothelial cells (second and third passage), on either a polystyrene or glass substrate, and exposed to a constant shear stress in the range of 8-16 dynes/cm^2 for a time period of anywhere from one to five hours. Although only preliminary results are available at this time, there has been a clear indication of the presence of a shear stress effect, but one which is highly dependent on substrate attachment.

D. Arterial Fluid Dynamics

The hot-film constant temperature anenometer and the pulsed ultrasonic Doppler system have provided for the first time insight into the temporal and spatial velocity variations present locally in the vascular system. Using such point velocity measurement systems, in vivo studies have been carried out in a variety of animals and in humans [24-28]. Furthermore, laser Doppler anemometry has been used in physical model studies [7]. From this combination of approaches, a much clearer picture of arterial fluid dynamics has emerged.

The classical problem used to explain the detailed characteristics of the flowing blood is that of Poiseuille flow. In this the flow is likened to that in a long, rigid, circular pipe with a constant flow rate. The velocity pattern or velocity profile, i.e. the variation in velocity as one moves radially across the lumen of the vessel from one wall to the other, is characterized as being parabolic. However, we now know that flow in the circulation system in general is not Poiseuillian in nature, i.e. it is not a fully developed viscous flow characterized by a parabolic profile, and there are many reasons for this. Primary among these is the unsteady nature of the flow in large arteries - the concept of Poiseuille flow strictly speaking only applies under steady state flow conditions. There is also the possibility of transition to turbulence [26], at least in the aorta. In the context of our interests here, however, one of the very major complexities of blood flow is that the vessels involved are often curved, and they branch repeatedly. This branching, asymmetric geometry quite naturally produces asymmetries in the velocity patterns, complicated secondary motions, entry phenomena and even possibly flow separation, all of which are far more difficult to analyze than the simple problem of steady state, fully developed Poiseuille flow [10].

These various aspects of arterial flow patterns are discussed in detail elsewhere [10,29]. However, associated with the local velocity pattern, there will be a fluid dynamic force exerted by the flowing blood on the arterial wall. The total stress imposed on the artery wall has both a normal component, i.e. the blood pressure, and a tangential component, i.e. the wall shear stress. Limited experimental data, as well as theoretical determinations, indicate peak wall shear stresses to be at the level of 100-200 dynes/cm^2 [10]. Although such shear stress levels are obviously much smaller than the stress due to normal pressure, it may well be that the arterial wall is designed to take a much higher normal stress. Furthermore, while in some general sense we believe we can characterize vascular regions as high shear or low shear as done in a previous section, there is a certain element of risk in doing this for in complicated flow regions the shear stress can vary considerably over just a very short distance.

E. Relationship Between Fluid Dynamics and the Arterial Wall

In this discussion of the case for fluid dynamics, it is important to examine the relationship of the detailed local flow pattern to endothelial changes and to the pattern of early lesions. Our knowledge is too limited - perhaps the interactions are just too complicated - to do this in a general way. However, it is possible to do this for the aorta and some of its major branch vessels. Figure 1 illustrates the flow pattern in the vicinity of an aortic branch vessel, for example, at an intercostal or where the coeliac artery branches off the aorta. The flow is unsteady and laminar, although possibly with turbulent bursts at peak systole in larger animals; and the velocity pro-

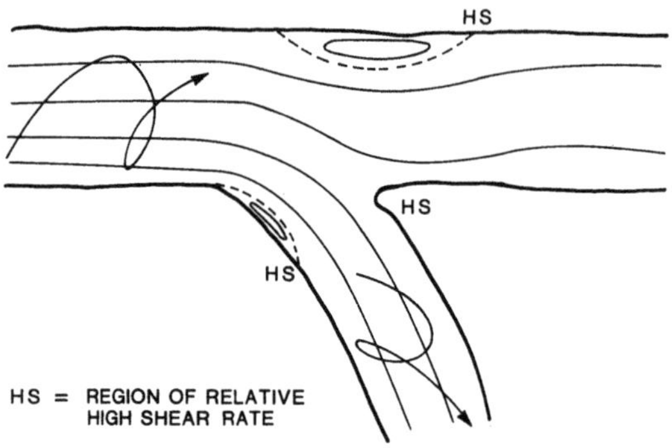

Fig. 1. Illustration of flow pattern in the region of an aortic branch vessel

file proximal to the branch point may be approximately symmetric, although not a parabolic Poiseuille flow due to the presence of entrance effects. However, immediately distal to the branch point the velocity profile will be asymmetric as streamlines move towards the wall from which the branch vessel emanates (this is illustrated in Figure 1). Further downstream the flow will adjust and tend to return to an approximately symmetric velocity profile. There will be secondary motions associated with the streamline curvature induced by the branching process and, in the thoracic aorta, due to the curvature of the aortic arch.

There is also the possibility of flow separation occurring. Although there are no in vivo observations to confirm this, the two most likely locations are illustrated in Figure 1. One of these is on the proximal side of the branching vessel as the flow turns to enter the branch, and the other location is on the wall opposite the flow divider. The likelihood of separation occurring depends for the first location on how sharp a turn the flow must negotiate and for the second location on the amount of flow drawn off through the branch. Even if flow separation does not occur, both of these regions would be characterized as generally low in shear, followed by a rapid increase in shear to produce a locally elevated wall shear stress. If flow separation does occur, this local increase in shear is due to the reattachment of the flow on the wall.

At the flow divider, there is a low shear, relatively high pressure stagnation region on the leading edge. However, once one is away from the leading edge region, there should be a relatively high shear both on the aortic surface as well as on the flow divider surface within the branch. This elevation in shear stress should persist for a distance on the order of a vessel diameter. Considering the iliac bifurcation for a moment, qualitatively the flow field would be similar to that just described. The leading edge of the flow divider would be a stagnation point region, and immediately distal on the flow divider surface the presence of a high shear stress should be expected. On the lateral walls, the shear will be low; however, there is a rapid increase in shear near the entrance to the daughter vessels which does not appear to be associated with the reattachment of a separated flow [30].

With regard to the endothelium, a picture of it also is beginning to emerge. First, it appears to be intact with desquamation occurring without denudation [31]. The endothelial cells are of an elongated character and appear to be aligned with the direction of the blood flow [22]. In regions of branching where the flow pattern may be quite complicated as just discussed and as is illustrated in Figure 1, the cells will not necessarily be aligned with the axis of the blood vessel. Furthermore, cells will be more rounded in some regions than in others. This appears to be related to local flow conditions and may be of some importance since the recent paper by Chazov et al. [32] suggests that atherosclerosis in humans is more likely to occur in regions having more rounded endothelial cells.

To be more specific about the shape of aortic endothelial cells in a region of branching such as illustrated in Figure 1, as one approaches the branch point from the proximal side, the cells on the surface from which the branch emanates will become increasingly more round. On the leading edge of the flow divider, the cells will also be of a generally round shape; however, here and in contrast to other locations, the pattern will be highly disorganized, demonstrating no particular preferred orientation, and presumably indicative of a stagnation region. On the aortic flow divider surface immediately distal to the leading edge, there is a region a few cell lengths long in which the cells are highly elongated, particularly when compared to the shorter, more rounded cells proximal to the branch point. On the aortic surface opposite the flow divider for the coeliac artery, which is a large branch vessel, there also is a region of more rounded cells. Also, in the region of the iliac bifurcation, endothelial cells appear to be more rounded on the outer, lateral walls and more elongated on the flow divider.

All of this taken together suggests that endothelial cells are more elongated in regions of high shear and more rounded in regions of low shear. Although this must be considered a preliminary assessment due to our still incomplete knowledge of the details of the flow field, the earlier noted results from in vitro studies of cultured endothelial cells exposed to a known flow condition support the existence of such a relationship between the shape of an endothelial cell and the level of wall shear stress.

With regard to the pattern of the disease in the adult human, Caro et al. [3] were the first to show that in fact lesions formed preferentially on the proximal side of aortic ostia. This would be a region believed to be low in shear, and the existence of a general pattern indicating low shear regions to be of higher predilection has been confirmed in other studies of the adult human [6] as well as in studies using the White Carneau pigeon. It is also consistent with the suggestion of Chazov et al. [32] indicating that regions of more rounded cells have a higher predilection for atherosclerosis since more rounded cells are believed to be associated with regions of low shear.

There are, however, two cautionary notes. First, the pattern of lesions in diet-induced atherosclerosis in animals is in general not in agreement with the above noted pattern [5]. The same may be said for the pattern of early lesions in fetuses, neonates, and infants [4]. Although this can possibly be explained on the basis that the lipid environment and thus the disease is not the same as in the adult human case, this remains to be further investigated. Secondly, although the evidence surveyed here provides an indictment for low shear regions as having the highest predilection for early lesions, it must be reemphasized that our knowledge is still too incomplete to say that with any great certainty. In fact whether it is shear, rapid changes in shear as reflected in the shear pulse amplitude, or some other related fluid dynamic event remains to be demonstrated in a more definitive fashion.

F. Concluding Discussion

All of the evidence for a fluid dynamic involvement presented here are in fact geometry related. This thus suggests that vascular geometry, and the way it influences the local detailed flow properties, may be the primary determinant of a hemodynamic effect on atherogenesis. Under such a hypothesis, whatever affects vascular geometry would alter the local, detailed flow characteristics and correspondingly influence endothelial morphology and function. For an individual subject, that geometry might be inherited, i.e. a family history or genetic effect, and as such one's geometry may in and of itself be a risk factor. Alternatively, a subject's vascular geometry may have been significantly altered by a risk factor such as hypertension. However, whatever the effect, the primary determinant of the pattern of atherogenesis would be vascular geometry, and whether or not an individual has an abnormal predilection for the disease would depend not only on biochemically related factors, but also on the exact nature of the geometry, e.g. the branching pattern and degree of curvature and/or tortuosity, and the resulting detailed flow characteristics.

A few years ago the prevailing view that if one could produce a statistically significant picture of the pattern of the disease and a correspondingly statistically significant picture of the flow pattern, then one would be in a position to draw conclusions on the relationship of hemodynamics to atherogenesis. However, it now appears that it is not the normal pattern of disease (in the sense of mean data representing the statistically normal situation) nor the normal flow pattern which are of interest. Rather, it is *deviations* from the *normal* which are important, and it is in this context that *vascular geometry* may be a *risk factor*. From this point of view, vascular geometry, whether inherited or as altered by the influence of a factor such as elevated blood pressure, would determine the local, detailed hemodynamics and the resulting influence on endothelial morphology and function. This, together with other, biochemically-related factors, would determine the focal nature of the disease. In this context, if one can understand the exact influence of vascular geometry on the local, detailed characteristics of arterial flows and the specific role of fluid dynamics in the biologic processes which lead to the genesis of atherosclerosis, then as vascular imaging techniques become more advanced, we may be able to tell an individual at an early age whether or not their vascular geometry puts them at an above normal risk for the development of atherosclerosis.

Acknowledgement. The authors thank S.A. Altobelli, C.G. Caro, C.J. Schwartz, and W.A. Seed for their contributions to the work and ideas reflected in this paper. The authors also are appreciative of the support provided by the National Science Foundation through Grant CME-80-01701.

References

1. Mitchell JRA, Schwartz CJ (1965) Arterial disease. Blackwell Scientific Publishing Ltd, Oxford
2. Montenegro MR, Eggen DA (1968) Topography of atherosclerosis in the coronary arteries. Lab Invest 18: 126
3. Caro CG, Fitz-Gerald JM, Schroter RC (1971) Atheroma and arterial wall shear: observation, correlation and proposal of a shear dependent mass transfer mechanism for atherogenesis. Proc R Soc Lond (Biol) 177: 109
4. Sinzinger H, Silberbauer K, Auerswald W (1980) Quantitative investigation of sudanophilic lesions around the aortic ostia of human fetuses, newborn, and children. Blood Vessels 17: 44

5. Cornhill JF, Roach MR (1976) A quantitative study of the localization of atherosclerotic lesions in the rabbit aorta. Atherosclerosis 23: 489
6. Kjaernes M, Svindland A, Walløe L, Wille SØ (1981) Localization of early atherosclerotic lesions in an arterial bifurcation in humans. Acta Path Microbiol Scand Sect A 89: 35
7. Friedmann MH, Hutchins GM, Bargeron C, Deters OJ, Mark FF (1981) Correlation of human arterial morphology with hemodynamic measurements in arterial casts. ASME J Biomech Engr 103: 204
8. Cornhill JF, Levesque JJ, Nerem RM (1980) Quantitative study of the localization of sudanophilic coeliac lesions in the White Carneau pigeon. Atherosclerosis 35: 103
9. Caro CG (1975) Mechanical factors in atherogenesis. In: Hwang NCH, Norman NA (eds) Cardiovascular flow dynamics and measurements. University Park Press, Baltimore, MD, p 473
10. Nerem RM (1981) Arterial fluid dynamics and interactions with vessel walls. In: Schwartz CJ (ed) Structure and function of the circulation, vol 2. Plenum Press, NY, p 719
11. Fry DL (1976) Hemodynamic forces in atherogenesis. In: Cerebrovascular diseases. Raven Press, NY, p 77
12. Velican D, Velican C (1981) Accelerated atherosclerosis in subjects with some minor deviations from the common type of distribution of human coronary arteries. Atherosclerosis 40: 309
13. Somer JB, Evans G, Schwartz CJ (1972) Influence of experimental aortic Evans blue uptake in vivo. Atherosclerosis 16: 127
14. Fry DL (1969) Certain histological and chemical responses of the vascular interface to acutely induced mechanical stress in the aorta of the dog. Circ Res 24: 93
15. Caro CG, Nerem RM (1973) Transport of ^{14}C-4-cholesterol between serum and wall in perfused dog common carotid artery. Circ Res 32: 187
16. Nerem RM, Mosberg AT, Shwerin WD (1975) Transendothelial transport of ^{131}I-albumin. Biorheology 12: 31
17. Chien S, Lee MML, Laufer LS, Handley DA, Weinbaum S, Caro CG, Usami S (1981) Effects of oscillatory mechanical disturbance on macromolecular uptake by arterial wall. Arteriosclerosis 1: 326
18. Thibault LE, Fry DL (1980) Hydrodynamically induced wall shear stress effects on Evans blue dye uptake. In: Nerem RM, Guyton JR (eds) Hemodynamics and the arterial wall. University of Houston, Houston, TX, p 34
19. Vaishnav RN, Patel DJ, Atabek HB, Deshpande MD, Plowman F, Vossoughi J Determination of the local erosion stress of the canine endothelium using a jet impingement method. ASME J Biomech Engr (to be published)
20. Languille BL (1981) Integrity of endothelium following exposure to high shear stress. In: 6th International Symposium on Atherosclerosis Abstracts. Berlin, FRG, June 13-17, p 616
21. Cornhill JF, Levesque MJ, Nerem RM (1980) The effect of arterial pressure on endothelial cell morphology. 2nd Austrian Atherosclerosis Conference, Vienna, April 7-12
22. Nerem RM, Levesque JJ, Cornhill JF (1981) Vascular endothelial morphology as an indicator of blood flow. ASME J Biomech Engr 103: 172
23. Dewey CF, Bussolari SR, Gimbrone MA, Davies PF (1981) The dynamic response of vascular endothelial cells to fluid shear stress. ASME J Biomech Engr 103: 177
24. Schultz DL, Tunstall-Pedoe DS, Lee GJ, Gunning AJ, Bellhouse BJ (1969) Velocity distribution and transition in the arterial system. In: Wolstenholme GEW, Knight J (eds) Circulatory and respiratory mass transport. Churchill, London, p 172
25. Seed WA, Wood NB (1971) Velocity patterns in the aorta. Cardiovascular Res 5: 319
26. Nerem RM, Seed WA, Wood NB (1972) An experimental study of the velocity distribution and transition to turbulence in the aorta. J Fluid Mech 52: 137
27. Nerem RM, Rumberger JA Jr, Gross DR, Muir WW, Geiger GL (1976) Hot-film coronary artery velocity measurements in horses. Cardiovasc Res 10: 301
28. Wells MK, Winter DC, Nelson AW, McCarthy TC (1977) Blood velocity patterns in coronary arteries. ASME J Biomech Engr 99: 26

29. Pedley TJ (1980) The fluid mechanics of large blood vessels. Cambridge Univ Press, Cambridge, England
30. Walburn FJ, Stein PD (1982) Shear rate at the wall in a symmetrically branched tube simulating the aortic bifurcation. Biorheology 19: 307
31. Schwartz SM, Gajdusek CM, Selden SC (1981) Vascular wall growth control: the role of the endothelium. Arteriosclerosis 1: 107
32. Chazov EI, Repin VS, Smirnov VN, Orekhov AN (1982) Intimal cell culture of atherosclerotic human aorta: a study of proliferation and lipoidosis. In: 6th International Symposium on Atherosclerosis Abstracts, Berlin, FRG, June 13-17, p 17

Discussion

FRIEDMAN: I was very happy to hear about your interest in geometry as a localizing factor. As part of our own experiments, we have been studying the unsteady flow of fluids through casts of arteries and examining the fluid dynamic phenomena that might occur. In these experiments we have not found any flow separation of consequence when using a fluid dynamically scaled pulsatile flow in the human aortic bifurcation. Now, in your experiments where you use the casts of the intercostal ostia to look at the images of the endothelial cells and you found more elongated cells distal to the ostium, could you make any estimate as how far distally one has to go before the cells return to approximately the same shape index that they had before the bifurcation, i.e. how extended is the effect of that new flow developing region at the lip? How far distally does it extend?

NEREM: Let me answer your question, but also respond to your comments. I was interested in your statement about flow separation and the fact that you do not see it. There has also been some work done by Fred Walburn and Paul Stein in Detroit, and they do not see any flow separation in the iliac region. What they do see, though, is a drop in shear rate and then a very rapid increase in shear as one enters into the parent vessels. They see a shear field that one might expect with flow separation, but they do not see flow separation itself. Thus, for example, when I say we see rounded endothelial cells opposite the coeliac artery, there may not be flow separation there, but there still may be some kind of an alteration in flow taking place such that you go through the same type of wall shear variation as one moves distally along the wall. With regard to your question, what we observed in the endothelial casts was a V-shaped region distal to the flow divider which was relatively short and on the order of about three cells in length, after which the cells returned to the normal pattern. This is the best quantitative number I can give you at this point.

SCHMID-SCHÖNBEIN: Is it not possible that those areas which naturally are exposed to stagnation point flow have the time of 25-40 years to cope with this and that this so-called cushion which develops should be looked into and viewed more as a mechanism for clearance. That is all we need to explain that this is not an area where atherosclerosis has developed, but it is an area where of course under all conditions we see higher endothelial turnover and higher uptake of high molecular weight substances. I think that, when we look at this process, we have to keep in mind that this is the unique thing about the arteries. There is no such thing as a flow divider in the veins. But on the other hand, there is a chance within the first three decades of our life to develop something which provides better lymphatic or better capillary clearance. Also, with regard to the area of so-called low shear, whatever the extent of it, whenever you have large pulsations, you can very easily get short instabilities which will affect the reattachment point. As we will hear later, platelets are ideally suited to attach to any wall provided that (1) they are stimulated and (2) the wall is accepting of it. So it could actually be that the question is not one of high shear versus low shear, but are there areas which are not prepared by life-long previous history to accept the changes associated with a moving stagnation point in a pulsatile flow?

CORNHILL: I think it is a very important point that you are making, i.e. the fact that it is extremely dangerous for us to suggest that just because an area is proximal or distal to a bifurcation, that it is an area of high or low wall shear stress. It is important to realize that since we are unable to measure these factors locally, we must be very careful. With regard to the question of changing geometry, if we have the change suggested by hypertension, we would in fact have a bifurcation which is modelled with flow from centered left to right and right to left and so these types of things are changing. I think that this even adds more to the possible danger of suggesting areas of high or low wall shear stress simply on the basis of being proximal or distal to a branch point.

NEREM: It may be that there is no flow separation for a normal geometry, but only as it becomes altered through hypertension such that the angle effectively gets reversed. Maybe there is even some separation on the flow divider itself as the flow comes around the corner.

BLACKSHEAR: When you look at the endothelium on the casts that you make, do you get any correlation between incidence of missing endothelium and age of the rabbits?

NEREM: We have looked at some older animals, and we did not see any endothelial changes.

GLAGOV: I just want to raise another point of caution which is that when you say that the animal models tend to have the atherosclerosis deposits just beyond the flow divider, it really depends upon the animal model.

NEREM: In the rabbit as usually used, there are very high serum cholesterol levels for short time intervals, and it is true that the atherosclerosis occurs just beyond the flow divider initially, but then it appears all around the ostium later anyhow.

GLAGOV: This is correct. However, in monkeys looked at for a short time with high cholesterol level, it occurs opposite the flow divider on the inlet. That is very common and only later does it occur beyond. So it depends upon the animal model and the duration and on the level of hypercholesterolemia, and you cannot just say that in the animal models it is beyond the flow divider.

NEREM: I accept that totally. I do feel that the lipid level that is used or that is present is a very important factor, and I do not quite understand yet what the relationship is. Certainly the kind of lipid levels involved in the monkey experiment is not the same as that used when we traumatically insult the system in the typical high cholesterol rabbit experiment.

S. SCHWARTZ: I think that the connection of rheology with the endothelium is interesting, but I would suggest that it might be worthwhile to think about what rheology means for what is deeper in the wall. If one goes back to the 1930s and 1940s, there were quite elegant studies done on angiogenesis which suggested that rheology was critical for the way which vessels develop. The Velicans and Wilbur Thomas have provided data that atherosclerotic lesions begin at branching sites because of developmental abnormalities that are here, entrapping smooth muscle cells in the intima. The branches may occur where they do because of flow, and much of the localization data that you show may be because of the effects of flow on the development of the vessels and the localization of smooth muscle cells in the vessel wall at places where the vessel has to continually be remodelled as it is developing a branch.

NEREM: I certainly do not think that we are talking about a disease of the endothelium, but I have taken the view, as you correctly pointed out, that the endothelium is the focus of the hemodynamic interaction. Although there may be things happening deeper within the wall, somehow the endothelium is involved in the sense that it is the mediator of the hemodynamic effect to lower levels. But perhaps we have put too much attention on the endothelium. The other thing I would comment on here is that we talk about endothelial cells being oriented by the flow. It may be that what is being oriented is the substrate material, and the endothelial cells then may just go along with that substrate. This may result in their being oriented with the flow, but the key thing is what is happening with the substrate material as opposed to the endothelial cell itself.

DEWEY: I want to refer to several comments that were made about the distribution of high and low shear in arterial bifurcations and perhaps to correct a small detail that Bob Nerem mentioned in one of his figures. There have been a number of studies of these bifurcations, as I am sure everybody is aware, and we will hear a whole session of them later today; but the thing that keeps striking me in all of these studies, including the ones that we have done, is that even though there may be separated flow (whether it is separated or not we will hear a little more about this afternoon) the direction of flow on the outside of these bifurcations changes dramatically. There are in general very low shear levels, and even though you may have an apparent reattachment region as Bob Nerem described, the magnitude of the shear there in quantitative terms is very low compared to what it is on the divider itself, simply because all of the flow is going down the inside. Very little is on the outside, and the shear stress is about an order or of magnitude less than the pre-bifurcation levels. We have made measurements in those kinds of geometries which suggest that, indeed, the shear is about a tenth of the magnitude in the indisturbed vessel, and then as you go downstream in one of the daughter vessels, it finally recovers to the regular level. So it really is very low even though you might say that around the reattachment point there would be a local maximum. The thing that is most striking is that, in the pulsatile case, both the actual direction and magnitude of the shear changes during the cardiac cycle for many endothelial cells, particularly those in the carotid sinus and around the edges of the carotid sinus. Several years ago there were some beautiful micrographs by W.W. Meyer that showed that for the carotid sinus, even in small children, there is a beautiful ring right around the sinus that indicates where there was serious compromise of the endothelium. I think this may be in fact the very region where the flow direction may change dramatically on each heart beat.

NEREM: Just one comment here. I agree with you that the shear is lower on the opposite wall, but I am not convinced that the absolute shear level is what is in question. It may just be that the relative variation in shear is what is important in a particular endothelial region.

SCHMID-SCHÖNBEIN: It is quite conceivable that even with the highest resolution electron microscopy, we have very little idea about the physiochemistry of a viscoelastic cell. I totally agree that the time changes, of course, are probably much more important than the absolute levels. However, without trying to interfere with some of the later presentations, it also is entirely possible that the cuboidal cell that we see on the flow dividers are not necessarily the sign of damage. They may very well be the result of good adaptation.

Flow and Atherogenesis in the Human Carotid Bifurcation

D. P. Giddens, C. K. Zarins, S. Glagov, B. K. Bharadvaj, and D. N. Ku

I. Introduction

Arterial branchings are of particular interest in the study of atherogenesis because of the frequent occurrence of atherosclerotic plaques in these sites. The bifurcation of the common carotid artery into the internal and external branches is an important example due to its frequent involvement with arterial disease and to the fact that transient ischemic episodes and cerebrovascular accidents are often associated with these extracranial lesions. Despite this well-recognized fact, little quantitative information is available concerning the precise distribution of carotid lesions or of their relationship to local hemodynamic factors. Inasmuch as hemodynamic behavior represents a mechanical factor which may create a favorable environment for plaque localization in bifurcations, we have undertaken a study to provide a detailed description of the flow field in the carotid bifurcation and of the distribution of early atherosclerotic lesions in humans. Previously, we have reported on flow visualization in a model of the human carotid bifurcation and on velocity and wall shear stress measurements in the model obtained with laser Doppler anemometry under steady flow conditions [1,2]. Recently, we have harvested human carotid arteries and studied the distribution of atherosclerotic lesions in these vessels. This paper discusses the relationship between the pathological and fluid dynamic findings.

II. Methods

A. Fluid Dynamics Experiments

A study of 57 angiograms from 22 adults between 34 and 77 years of age and 67 angiograms from 50 youths below 18 years of age was performed to derive an average carotid bifurcation. Bi-planar angiograms were always used to obtain data for the branch angle. Two models were constructed, one from blown glass for flow visualization and one from machined plexiglass for the laser Doppler anemometer (LDA) studies. Figure 1 gives the configuration for the two models and indicates the sites for which measurements were taken from the angiograms. Table 1 gives the corresponding results for these measurements.

Data available on blood flow rate through the carotid vessels is extremely variable [3-5]. We selected values of 500, 350 and 150 ml/min as representative of the mean blood flow rate through the common, internal and external carotid arteries. Using the common carotid vessel as an upstream reference condition with 8 mm for the diameter and assuming 0.035 poise as the viscosity for blood, the value of mean Reynolds number (Re) is 380. Inasmuch as the flow is pulsatile, peak values of Re can be as high as 1200. Consequently, in the experimental studies Reynolds numbers of 400, 800 and 1200 were employed and flow division ratios for the internal:external branches of 60:40, 70:30 and 80:20 were utilized to cover the physiological range of parameters.

Fluid Dynamics as a Localizing Factor for Atherosclerosis
Editors: G. Schettler et al.

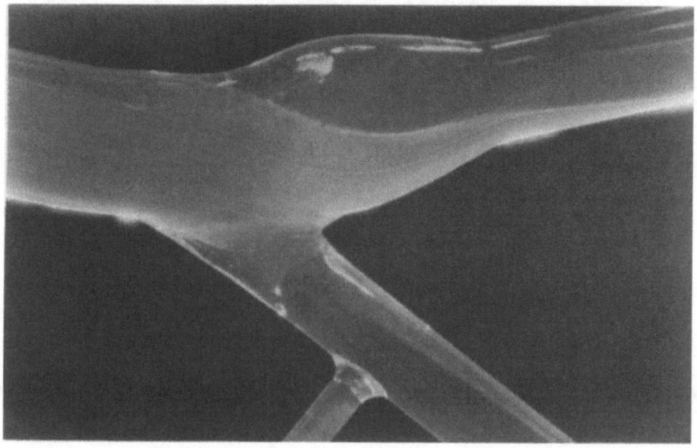

Fig. 1. Hydrogen bubble visualization of flow separation in the carotid sinus (Re = 400, 70:30 flow division)

Flow visualization was achieved by dye injection and hydrogen bubble techniques, and fluid velocities were measured with an LDA system (DISA, Mark II, single component). Water/glycerin mixtures were employed to assist in achieving appropriate Reynolds numbers. The common carotid vessel of the model had an internal diameter of 31 mm and an inlet length of 2.7 m to allow fully developed upstream conditions. Flow rates in the daughter branches were monitored with an electromagnetic flowmeter. Additional details of the experimental arrangement, along with a discussion of the relative importance of steady flow, rigid walls and a Newtonian fluid are all discussed in previous reports [1,2]. Wall shear stresses were estimated from near-wall velocity profile measurements (three velocity data points obtained within 0.5 mm of the wall). Corrections were made in the velocity data for index of refraction and curved wall effects. The LDA velocity data, obtained from the Doppler signal with a frequency tracker, were digitized and processed on a Hewlett-Packard HP-5451B Fourier Analyzer System.

B. Morphology Studies

Twelve carotid bifurcations were obtained at autopsy from subjects aged 27 to 73 years who had no history of symptomatic cerebrovascular disease. The branches of the external carotid artery were ligated, and the internal and common carotid vessels were cannulated. The bifurcation was then warmed to 37°C and perfused to a 100 mm Hg pressure level before final perfusion and immersion employing warm buffered formalin. The intraluminal pressure was maintained during this procedure. The fixed vessels were next perfused with a radiopaque material at 37°C and 100 mm Hg pressure and allowed to cool to room temperature. Angiograms of the bifurcations were taken at several projections and measurements, corresponding to those obtained in the model study, were scaled from the results. No carotid bifurcations which showed evidence of luminal stenosis on X-ray results were included in the study.

Histologic examination was performed on transverse sections at five standard locations, corresponding as closely as possible to locations at which velocity profiles were obtained in the plexiglass model. The microscopic sections were projected onto a digitizing plate; and contours of the lumen, internal elastic lamina, and outer boundary of the media were identified and digitized with a microcomputer. Intimal and medial thicknesses were determined at 15° intervals around the vessel circumference. The 0° reference point in the internal carotid artery was taken along the wall forming the flow divider. Consequently, 180°

was located at the outer wall of the sinus, while 90° and 270° corresponded to the midpoints of the anterior and posterior halves of the vessel. The percent stenosis was calculated at each section by dividing the area occupied by the intima by that contained interior to the internal elastic lamina.

III. Results

A. Flow Studies

The flow visualization studies demonstrated that very complex fluid dynamic patterns occurred in the model. Of particular significance was the fact that flow separation occurred in the carotid sinus at all conditions studied (Figure 1). High velocity fluid was directed towards the flow divider, while the sinus was filled with slower-moving helical patterns (Figure 2).

Axial velocity profiles were measured in the plane of the bifurcation at the various flow conditions. An overview of the results for Re = 400 and a 70:30 flow division is shown in Figure 3. The large region

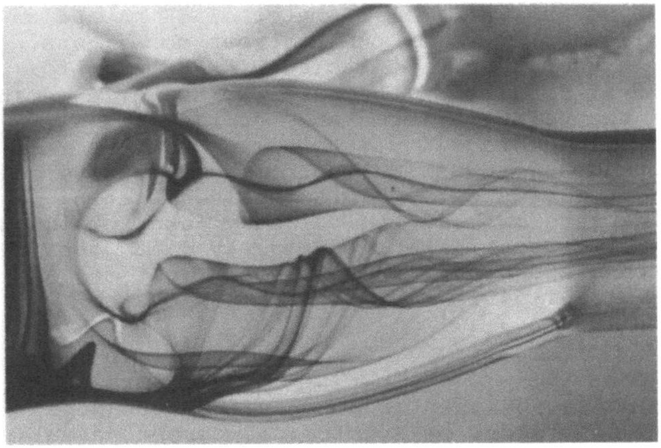

Fig. 2. Dye visualization illustrating presence of helical secondary flow patterns in the carotid sinus (Re = 1200, 70:30 flow division)

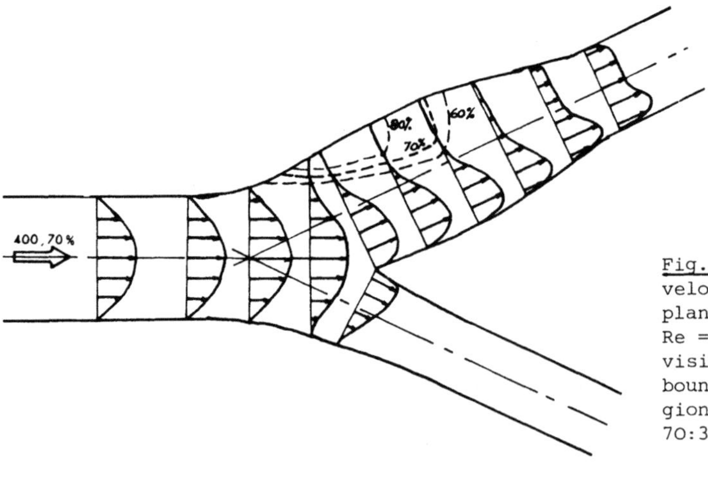

Fig. 3. Overview of axial velocity profiles in the plane of the bifurcation for Re = 400 and 70:30 flow division. Dashed lines show boundary of reverse flow region for Re = 400 and 60:40, 70:30, 80:20 flow divisions

of separated flow, leading to a zone of relatively low wall shear stress, may be seen along the outer wall. Figure 4 presents a series of axial velocity profiles at these same stations, but with Re = 400 and a 60:40 flow division. Although the peak velocities in these profiles are comparable in magnitude with the centerline velocity in the common carotid vessel, the peak has shifted very close to the inner wall (flow divider), resulting in relatively large wall shear stresses. Since flow was symmetric above and below the plane of the bifurcation, the wall shear stresses along 0° and 180° could be estimated from the axial velocity profiles alone.

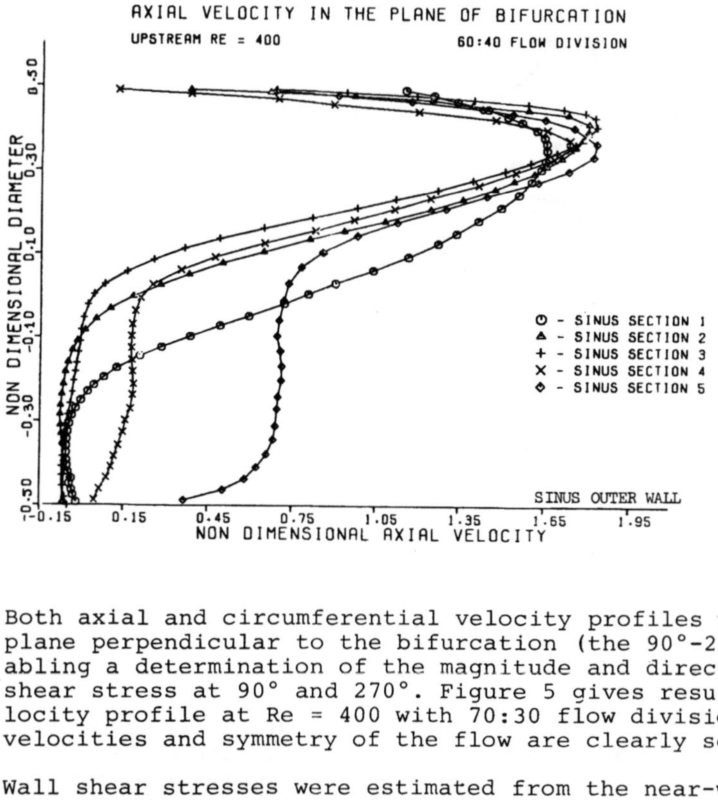

Fig. 4. Axial velocity profiles in the carotid sinus, taken in the plane of the bifurcation (Re = 400, 60:40 flow division)

Both axial and circumferential velocity profiles were measured in the plane perpendicular to the bifurcation (the 90°-270° plane), thus enabling a determination of the magnitude and direction of the wall shear stress at 90° and 270°. Figure 5 gives results for the axial velocity profile at Re = 400 with 70:30 flow division. The off-axis peak velocities and symmetry of the flow are clearly seen in these data.

Wall shear stresses were estimated from the near-wall velocity gradient. Of particular interest was the behavior of this quantity along the outer wall of the sinus. Figure 6 presents the variation of τ_w for 70:30 flow division and at Re = 400, 800 and 1200. The variation of the dimensionless form of τ_w was observed to be fairly insensitive to Reynolds number over the range studied. Negative values correspond to regions of reverse flow, and a dimensionless value of 1.0 corresponds to the fully developed Poiseuille value. Additional results on the wall shear stress variation for other flow conditions may be found in [6].

B. Morphology Studies

The linear dimensions for the model carotid bifurcation were all obtained as a ratio to the diameter of the common carotid artery. Based upon data available in the literature [7], we took this value to be 8 mm. For the carotid vessels harvested for morphological study, the average diameter of the common carotid artery was 6.1 mm. Consequently,

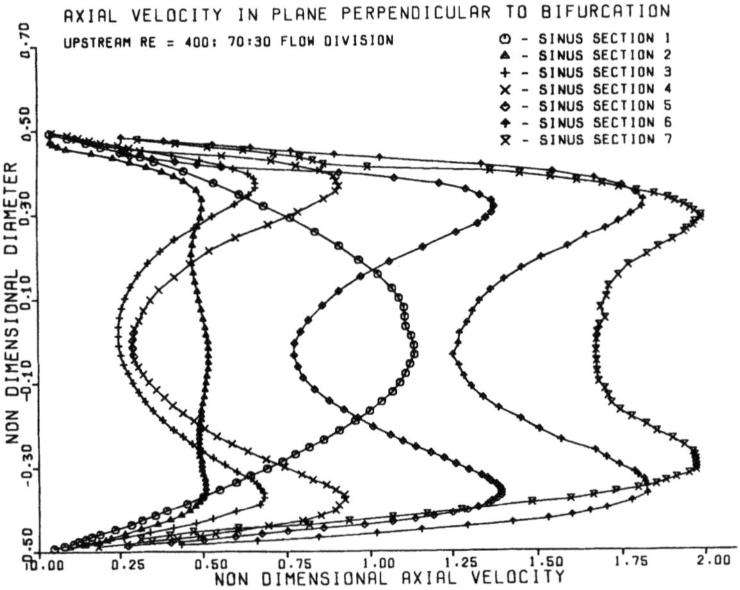

Fig. 5. Axial velocity profiles in the carotid sinus, taken in a plane perpendicular to the plane of the bifurcation (Re = 400, 70:30 flow division)

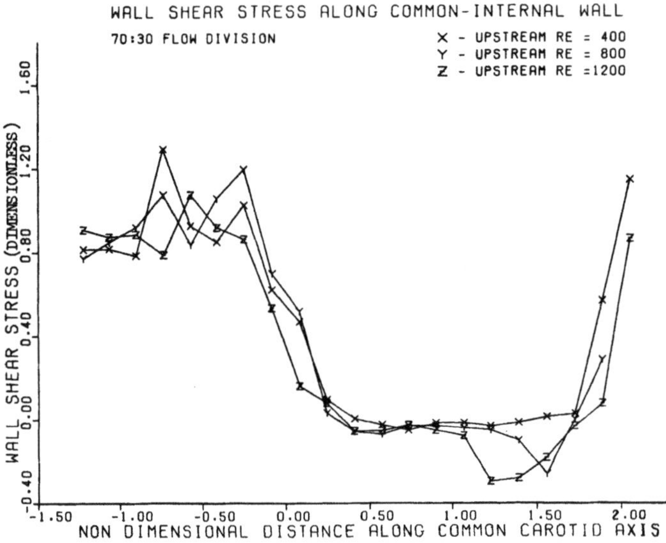

Fig. 6. Variation of wall shear stress along the common-internal wall for different upstream Reynolds numbers and 70:30 flow division

when scaling our fluid dynamic results for application to the human case, we employed this value (6.1) rather than 8 mm. A comparison of the average dimensions for the specimens studied with the scaled model dimensions (which were obtained approximately 2 years earlier from the angiographic studies) is given in Table 1. The agreement is remarkedly good, the greatest discrepancies being in the length of the sinus bulb and the diameter of the distal internal carotid artery.

It was interesting to note that, despite the fact that postmortem angiographic evaluation of the harvested vessels gave no indication of stenoses, all the subjects presented localized intimal thickening

Table 1. Comparison of average dimensions among human specimens studied with the scaled dimensions of the plexiglass model bifurcation used for in vitro flow experiments

	Human carotid bifurcation (N=12)	Model bifurcation (Scaled dimensions)
Lumen diameter (mm)		
(A) Common carotid	6.1 ± 0.2	6.1
(B) Proximal internal carotid	6.0 ± 0.4	6.3
(C) Midpoint of carotid sinus	6.2 ± 0.3	6.8
(D) Distal internal carotid	3.5 ± 0.2	4.2
(E) External carotid	4.0 ± 0.2	4.2
Length of int. carotid bulb (mm)	19.0 ± 1.0	13.0
Bifurcation angle (degrees)	46 ± 7°	50°

Mean ± S.E.M.

or plaque formation upon histological examinations. The morphometric quantitation studies demonstrated that intimal thickening and plaque formation tended to localize in the carotid sinus and, more specifically, in the vicinity of the outer wall of the sinus. Figure 7 presents the X-ray image and a vessel cross-section from one of the subjects. The plaque localization towards the outer wall was typical among the results.

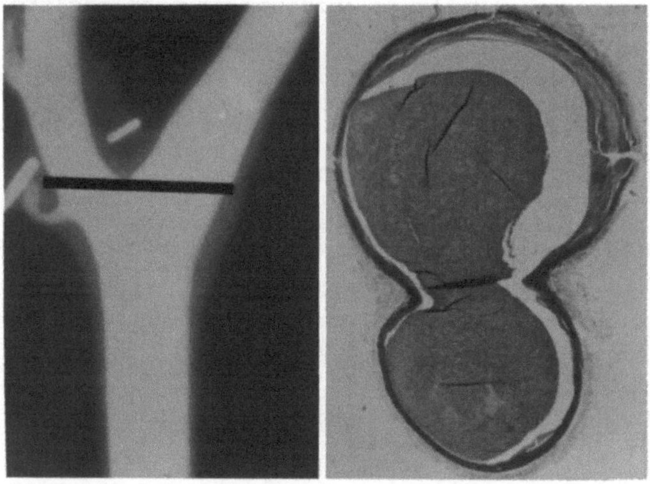

Fig. 7. Example of a human carotid bifurcation harvested and perfusion-fixed for study: angiogram taken perpendicular to bifurcation plane and cross-section taken just at flow divider

Table 2a presents the intimal and medial thicknesses for the mid-section of the carotid sinus at 0°, 90°, 180° and 270°. Although there is no significant difference in the medial thickness as a function of the polar angle, there is a strong tendency towards intimal thickening near 190°, the outer wall. Table 2b gives the corresponding wall shear stress data obtained from the in vitro model study after scaling the results to the dimensions of the human artery. The values are given in dynes/cm^2 and are presented for Re = 400, 800 and 1200. There is a strong correlation between the areas of low wall shear stress and the region over which intimal thickening was more prominent.

Table 2. Intimal and medial thicknesses for the midsection of the carotid sinus at 0°, 90°, 180° and 270°, and corresponding wall shear stress measurements obtained by scaling model data to the human carotid bifurcation. Site of comparison is the midpoint of the carotid sinus

	Polar coordinate position			
	0°	90°	180°	270°
Intimal thickness (mm)	0.14 ± 0.05	0.47 ± 0.13	0.63 ± 0.17	0.37 ± 0.10
Medial thickness (mm)	0.21 ± 0.03	0.21 ± 0.02	0.32 ± 0.03	0.26 ± 0.02
Wall shear stress (dynes/cm^2)				
Re = 400	31	--	0	--
Re = 800	120	46	-3	46
Re = 1200	160	93	-6	93.

IV. Discussion

We are continuing to harvest human carotid bifurcations and are refining the quantitation of the morphometric analysis. However, results to date indicate a very strong relationship between regions of low velocity and low wall shear stress in the model internal carotid branch and the localization of intimal thickening and plaque in human specimens. The flow divider, which tends to experience relatively high wall shear stresses throughout the entire cardiac cycle, has a very strong tendency to be spared from disease. Similarly, in the segment of the internal carotid artery distal to the sinus, where the vessel narrows and flow accelerates, there is virtually no evidence of atherosclerotic disease.

The maximum intimal thickness occurred at 161° in the proximal segment of the carotid sinus. Our model was symmetric about the plane of the bifurcation and did not possess the additional gentle curvature characteristic of the human anatomy. Such geometric differences could perhaps be expected to lead to asymmetric plaque localization. Certainly, the in vitro flow studies clearly demonstrate that the flow field is quite complex and would be sensitive to changes in geometry. It is expected that the branch angle would be a prominent factor in dictating the flow division. Another factor which tended to affect the results appreciably was the flow division ratio. As the flow into the internal branch decreases, the region of separation increases noticeably. The pulsatile nature of the actual carotid flow may also play a role in the localization process, inasmuch as the flow is not truly quasi-steady at frequency parameters typical of the carotid artery. We do not expect, however, a great change in the strong correlation established between the present steady flow results and plaque localization.

In summary, our results demonstrate that intimal thickening and plaque formation in the human carotid bifurcation tend to localize in regions of low velocity and low wall shear stress. There is also significant plaque development in regions of the bifurcation which experience large changes in direction of the wall shear stress vector, i.e., the side walls of the sinus. We consistently saw sparing of disease in regions of relatively high velocity and moderate to high wall shear stress. Although these results do not establish a specific mechanism for lipid deposition, they clearly contradict the high shear stress/endothelial injury hypothesis for atherogenesis and tend to support the hypothesis that mass transport through normal endothelia may play an initiating role.

References

1. Bharadvaj BK, Mabon RF, Giddens DP (1982) Steady flow in a model of the human carotid bifurcation. Part I - Flow visualization. J Biomech 15: 349-362
2. Bharadvaj BK, Mabon RF, Giddens DP (1982) Steady flow in a model of the human carotid bifurcation. Part II - Laser-Doppler anemometer measurements. J Biomech 15: 363-378
3. Kristiansen K, Krog J (1962) Electromagnetic studies on the blood flow through the carotid system in man. Neurology 12: 20-22
4. Tindall GT, et al. (1962) Studies on carotid artery flow and pressure. J Neurosurg 19: 917
5. Cordell RA, et al. (1968) Clinical blood flow measurement. Evaluation of a current method. Ann Surg 167: 779-785
6. Balasubramanian K (1979) An experimental investigation of steady flow at an arterial bifurcation. Ph.D. Thesis, Georgia Institute of Technology
7. Olson RM (1974) Human carotid artery wall thickness, diameter, and blood flow by a non-invasive technique. J Appl Physiol 36: 955-960

Discussion

CORNHILL: Most of the data you presented were on the internal carotid. In the external carotid you referred very briefly to the exactly opposite effect. You saw a two- to three-fold increase in intimal thickness, at least on the flow divider. Can you comment on that? Can you give us any insight into what appears to be another difficult situation for us to explain hemodynamically?

GIDDENS: It is puzzling, however, if you want to stick to strictly a fluid dynamics interpretation, one possible explanation might be as follows. The flow division during the pulsatile cycle is not constant between internal and external carotids. As you know, the internal carotid feeds a low resistance bed, and in the external carotid, the flow during diastole most of the time goes to zero or in some cases even goes negative. So there are times during the cycle in which flow into the external carotid would be extremely low in the human. And as you well know, if you reduce the flow rate into that external branch, then you tend towards separation and more complicated flow patterns. One very speculative possibility is that in the human case these types of flow phenomena, e.g. flow separation and rather complicated helical flow patterns, would tend to exist in the external carotid artery during parts of the cycle.

CORNHILL: I think this is another very good example of what Dr. Glagov was talking about earlier where you have these apparent differences, not even within a species, within a given bifurcation.

BLACKSHEAR: When you looked at the occurrence of lesions in human material and contrasted it to the different properties of the arteries, did you look at the medial thickness in the carotid sinus and compare the distribution of lesions with thickness?

GIDDENS: Yes, the medial thickness was looked at, and there did not seem to be much direct correlation. There was a slight increase in the medial thickness in the proximal internal at the 180° point. This was also the point where there was considerable intimal thickening. Thus at that particular site there was some correspondence between medial thickness and intimal thickness, although the correspondence was not strong. At the other sites there was very little variation in the medial thickness.

Pulsating Flow in Distensible Models of Vascular Branches

D. Liepsch, S. Moravec, and R. Zimmer

Summary

Deposits and blockages often occur at bends and bifurcations of human arteries. In addition to biochemical factors, haemodynamics also plays a major role in arteriosclerosis. The haemodynamic factors - rate of flow, pressure, flow behaviour, shear stress on the walls and in the whirl zones - are therefore of particular significance to the formation mechanisms.

Of these, the shear stress on the edge of the whirl zones and flow-deprived areas are of special interest [1,6,7,8,9,10]. The duration for which the particles remain in these zones is also a decisive factor [2]. High and low shear stresses are believed to be responsible for the adhesion and deposition of thrombocytes [4,11]. The present study demonstrates velocity measurements in rigid and elastic models under pulsating flow conditions.

Experimental Methods

In order to investigate arterial blood circulation experimentally a hydrodynamic model is required. This must comply with the following conditions:

- geometrical similarity
- virtually identical flow behaviour of the fluid
- same hydrodynamic conditions (volume pulse, peripherical resistance)
- similar wall elasticity

The following experimental steps were carried out:

- under stationary flow conditions
- under pulsating flow conditions

using true-to-scale, rigid polyester resin models and subsequently using elastic silicon rubber models [5].

The models were prepared from plastic casts of human renal, leg and coronary arteries. The fluid passed through the models were firstly distilled water, then a sugar solution and lastly an aqueous polyacrylamide solution with the addition of dimethylsulphoxide. The latter mixture has a very similar flow behaviour to blood.

The selected volume flow ratios correspond to the physiological velocity of blood in the arteries.

Flow Indicators

In order to carry out a qualitative observation of the flow patterns in the models, different birefrigient solutions were passed through the models to localize the positions at which the flow was deflected from the wall, causing whirl zones. The observations were made using

Fluid Dynamics as a Localizing Factor for Atherosclerosis
Editors: G. Schettler et al.

photoelastic apparatus under stationary and pulsating flow conditions. In addition, the flow pattern was made visible by dyeing individual stream lines. The latter method is only applicable to stationary flow conditions.

Once the disturbed zones have been located, the flow velocity is measured at the disturbed zones immediately before and after the bifurcation.

Velocity Measurements with a Laser-Doppler-Anemometer

The principle of the laser-Doppler anemometer is well-known. The non-contact method of measuring allows velocities in tubes with a small diameter to be recorded. The experimental setup has already been described many times [3]. By using a 1.8 m run-in tube (L/D = 180), the flow is stabilized by the time it reaches the incorporated bifurcated model. The pressure gradient is measured using two inductive pressure sensors. The pulsating flow is generated by means of a piston pump. This superimposes a sinusoidal oscillation on the stationary flow.

The previous experimental setup enabled only one velocity component to be measured at a time, for example in the main stream parallel to the central axis of the tube and then perpendicular to the central axis. Investigations are currently being made into the possibility of measuring two velocity components simultaneously.

Results

Pulsating Laminar Flow

If an oscillating flow is superimposed on a stationary flow, this produces a pulsating flow. Figure 1 shows the recorded velocity distribution of a pulsating flow in the main branch of a 35° bifurcation [12]. One rigid model is compared with two elastic silicon rubber models with a wall thickness of 1 and 2 mm. The flow medium employed was a calcium chloride solution with a dynamic velocity η = 2.85 mPas at 23 °C and a density ρ = 1280 kg/m^3. This medium has the same refraction index as silicon rubber. Therefore the laser beam is not deflected when passing through the medium and the model. The flow characteristics were represented by the amplitude Amp of the differential pressure dp/dx, the Wormersley parameter $\alpha = R\ \omega/\nu$, where R radius of the tube, in figure 0.004 m, ω angular velocity $\omega = 2\pi f = 2\pi \times 1.4 = 8.8$, and ν kinematic viscosity 2.2×10^{-6} m^2/s, and the mean Reynolds number of the volume flow ratio $\dot{V}$ averaged over a period.

The pressure fluctuations were selected to correspond to the human arterial system (135/85 mm Hg). The numbers of the abscissa denote the distance from the bifurcation and the measuring plane. The velocity fluctuations at various phase angles ωt are applied on the ordinates. The annular effect at the tube wall at measuring point zero is very apparent. The silicon bifurcation with 2 mm thick walls shows a similar velocity pattern in the vicinity of the wall to that of the rigid tube. With a wall thickness of 1 mm a sinusoidal pattern was obtained at the measuring points r/R = 0.72 in the vicinity of the wall. Major differences in the velocity distribution only occur with phase angles of 90° - 180°. At other phase angles ωt, there is a constant velocity across the entire diameter of the tube. At these phase angles the negative velocities of the disturbed zones are reduced.

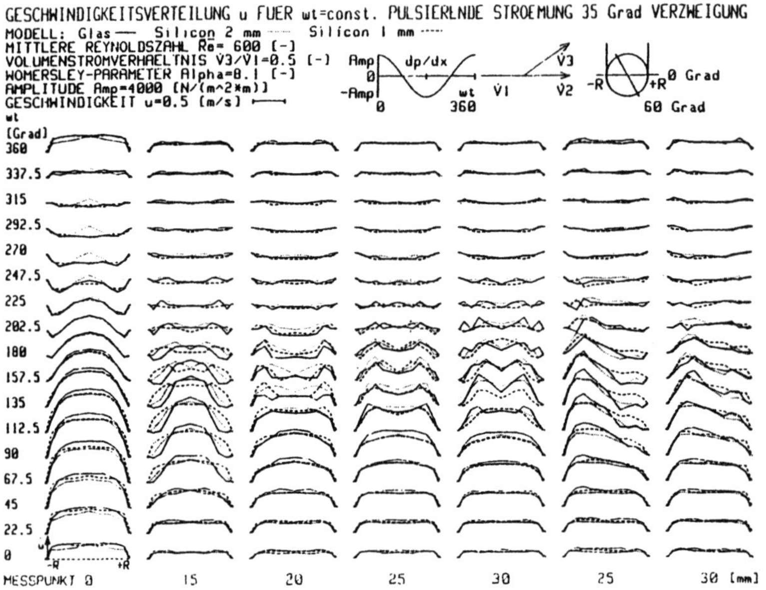

Fig. 1. 35°-bifurcation in a leg artery model. Comparison of velocity measurements in a rigid glass model and two elastic silicon rubber models under pulsating flow conditions

In the rigid model, counter-movement occurs between the positive main stream and the negative reverse flow in the disturbed zones. The counter-movement decreases as the elasticity of the wall of the model increases. In other words, in an elastic model the main stream oscillates in phase with the whirl zone, whereas in rigid models a phase shift occurs. This means that the reverse flow is increased. In models with elastic walls there is a continuous small forward movement. In models with stiffened or rigid walls, zones with reverse flow or on forward movement are formed. Particles in these locations oscillate at a point.

Pulsating Flow with a Non-Newton Fluid

If the experiments are carried out using a non-Newton fluid, even greater reverse flow is obtained after bifurcations. Figure 2 shows a comparison between a glycerine/water solution and a Separan solution in a renal artery bifurcation. 5 mm after the bifurcation it can be clearly seen how the reverse flow zones in the vicinity of the wall are more pronounced in the case of non-Newton fluid, virtually throughout the entire phase $\omega t = 0° - 360°$.

Conclusions

As demonstrated by earlier investigations on bifurcations, the stream is deflected from the wall in stationary flow conditions once the specific volume flow ratio of the secondary branch to the main stream $\dot{V}_3/\dot{V}_1$ exceeds 0.24. In each case this disturbed zone is located opposite the secondary branch. If the volume flow ratio is increased, a second disturbed zone is formed on the same side as the secondary branch, further downstream. Reverse flow occurs at these positions.

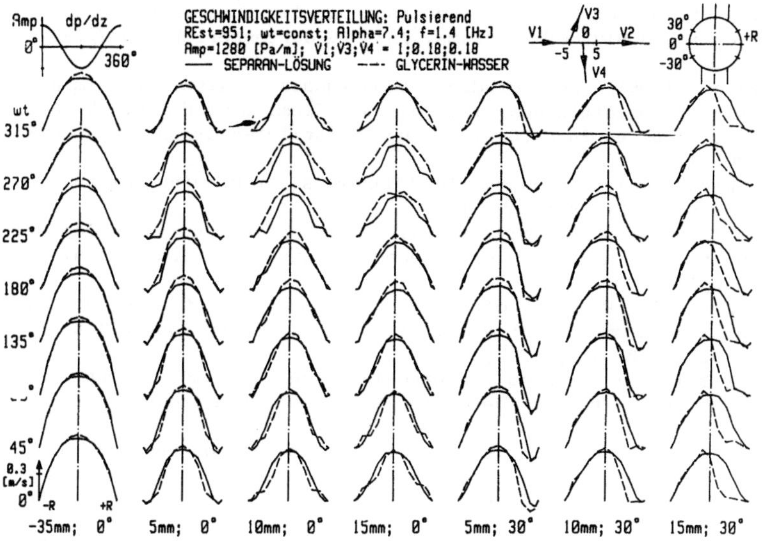

Fig. 2. Glass model of a renal artery model with Separan solution and glycerine/water solution as flow medium under pulsating flow conditions

In the case of pulsating flow through rigid models, the reverse flow areas of the bifurcation become accentuated in the main branch. High shear stresses are observed on the boundary between whirl zones and the main stream, since the negative velocity of the reverse flow and the positive velocity of the main stream oscillates in phase but in opposite directions. The shear forces can increase up to 60 times Hagen-Poiseuille flow (τ_{max} = 10 N/m^2). Also due to the high velocity, shear stress at the wall of the secondary branch may increase by 3 to 6 times. When particles collide with the wall at an estimated blood velocity of 0.2 m/s, pressures of up to approximately 40 N/m^2 can occur.

Where walls are elastic, these shear stress values are reduced by about 25 %.

If a substance with a non-Newton fluid pattern is used instead of a medium with a Newton flow pattern, the reverse flow is accentuated in the same way. Until now, it was assumed that a Newton flow pattern occurs in arteries with a diameter greater than 1 mm, and this is in fact true for arteries without bifurcations. But after bifurcations whirl zones occur and the flow parallel to the central axis is disturbed. Therefore the non-Newton flow behaviour of blood must not be ignored in the case of larger arteries. The whirl zones observed correspond to the predilection areas in human arteries.

References

1. Geigy Documenta, Thrombose in neuem Licht. Firmenschrift 1978
2. Karino T, Kwong HM, Goldsmith HL (1979) Particle flow behaviour in models of branching vessels: F. vortices in 90°-T-junctions. Biorheology 16: 231-248
3. Liepsch D, Moravec St, Zimmer R (1981) Einfluß der Hämodynamik auf Gefäßveränderungen. Biomed Technik 26: 115-122

4. Liepsch D, Moravec St, Zimmer R (1978) Untersuchung der Strömungsvorgänge in Modellen des menschlichen Blutgefäßsystems. In: DFG-Jahresbericht 1977/78. Nimbach, München
5. Moravec St, Liepsch D (1981) Geschwindigkeitsmessungen an einer dreidimensionalen Verzweigung mit viskoelastischen Fluiden als Medium. Biomed Technik 26: 119
6. Naumann A (1975) Strömungen in natürlichen und künstlichen Organen und Gefäßen. Klin Wschr 53: 1007-1019
7. Nerem RM, Rooz Elkana (1981) Coronary fluid dynamics. Proceedings of the Fourth International Congress of Biorheology, Tokyo
8. Niimi H, Yamakawa T, Hanai S, Minamiyama M (1981) Fine structure of blood flow near a wall in relation to atherogenesis. Recent Advances in Cardiovascular Disease, vol 2 [Suppl] 89-91
9. Schmid-Schönbein H (1981) Fluidity of blood in microvessels: Consequences of red cell behaviour and vasomotor activity. Proceedings of the Fourth International Congress of Biorheology, Tokyo
10. Talukder N, Reuter B (1981) Messung von Wandeinflüssen auf die Strömung makromolekularer Flüssigkeiten mit Hilfe des Laser-Differential-Mikroanemometers. Biomed Technik 26: 120
11. Wurzinger LJ (1979) Hydrodynamisch induzierte Plättchenablagerungen an Glasmodellen verzweigter Gefäßabschnitte und Speziesunterschiede im Plättchenaggregationsverhalten von Mensch, Rind, Schwein, Schaf, Hund, Kaninchen und Truthahn. Diss RWTH Aachen
12. Zimmer R, Liepsch D, Moravec St (1980) Geschwindigkeitsmessung mit einem Laser-Doppler-Anemometer in einer starren und elastischen 35°-Verzweigung. Biomed Technik 25: 368

Discussion

SEED: Can you tell us how the elastic properties of your models compare with the elastic properties of real arteries?

LIEPSCH: Yes, you can compare the models. We have the same elastic model, but not the hysteresis as in human arteries.

DEWEY: You probably have taken all of the most difficult things, including mechanics, and done them all at once. Not only do you have pulsatile flow and real geometry, but you also have a non-Newtonian fluid. It thus is very hard, obviously, to go through all of these data and try to understand what was caused by what; there are so many things going on at the same time. However, one of the things that was surprising to me, and also disturbing, was that with the use of the polyacrylamide non-Newtonian fluid, there was a difference of a factor of three in the shear stress. I really do not quite understand that. Furthermore, it is not clear to me that the non-Newtonian behavior of your polyacrylamide solution is going to be the same as that of blood, particularly when you have unsteady flow.

LIEPSCH: We must decide between the elastic and the other components, and what is the same for the polyacrylamide with the dimethylsulphoxide is only the elasticity of the medium. The visco-elasticity is not the same. We have a very small range where we have the same visco-elasticity, but not over the entire shear stress range. There is no medium which can simulate the entire shear stress range of blood.

Coronary Artery Geometry and Its Fluid Mechanical Implications

R. M. Nerem and W. A. Seed

Abstract

Detailed measurements of dimensions and branching geometry were obtained from 19 post-mortem casts of normal coronary arteries. The casts were prepared by injection of quick-setting resin at physiologic pressures.

Diameters of the major vessels scaled well to aortic diameter. The left coronary artery divided into its two branches after a variable distance, with an area expansion. The geometry of this bifurcation, and of other branches downstream, showed considerable variation. The right coronary artery also showed a variable pattern of branching, but had a constant curvature for the major part of its length.

Estimated Reynolds numbers were in the range 70-170, and mean wall shear-stress 12-42 dynes/cm, at both upstream and downstream sites in all major vessels. Comparison with known patterns of early atheroma suggest that if fluid mechanic determinants of atherogenesis operate in these vessels, they are determined on a very localized basis by the anatomy.

Introduction

It has long been established that clinically significant atherosclerotic lesions tend to occur at predictable sites in the circulation [1]. Recently, detailed studies of arteries obtained at post-mortem from young subjects have demonstrated in several vascular beds that this is also true for the early lesions of atherosclerosis [2,3,4,5]. Thus in the coronary arteries [3] very early fatty disease, like clinically important advanced disease, tends to be concentrated proximally in all major branches. Furthermore, there is an asymmetric distribution around the circumference of each vessel. Such observations, suggesting some topographic determinant of disease, are consistent in a general way with hypotheses which link the evolution of atheroma with fluid mechanic mechanisms [6,7]. However, the detailed geometric information which is a necessary background to the analysis of local fluid mechanics is not available. We have therefore attempted to define the anatomy of the human coronary arteries in terms likely to be relevant to local hemodynamics.

Methods

Post-mortem luminal casts of the coronary arteries were prepared from the hearts of young (16-40) subjects dying from accidental or non-cardiovascular natural causes. In each case, a perfusing cannula was tied into the ascending aorta, balloons were inserted into right and left ventricles and filled with fluid to physiologic volumes, and the heart suspended in a bath of isotonic saline. The perfusing cannula was supplied with isotonic saline from an elevated reservoir arranged to maintain perfusing pressure at 100 mm Hg. When the saline supply was

Fluid Dynamics as a Localizing Factor for Atherosclerosis
Editors: G. Schettler et al.

turned on, the aortic valve remained closed, and the coronary arteries were perfused from the ascending aorta at physiologic pressure. After flushing, a quick setting liquid resin was injected via a side arm of the perfusion cannula, at a rate sufficient to replace the saline entering the coronary arteries; thus the vessels were perfused with resin at physiologic pressure, whilst the main cardiac chambers contained physiologic volumes so that the shape of the heart was maintained. When the resin gelled, the cannula was clamped off and the resin allowed to harden for 24 hours. This technique fills the coronary arterial tree down to branches of 0.1 mm or less in diameter (Fig. 1). In the present study, only the epicardial vessels of the coronary arterial tree were examined. The epicardial part of the cast, still attached to the cast of the ascending aorta, was dissected from the heart and removed intact. Technically satisfactory casts were obtained from 19 hearts.

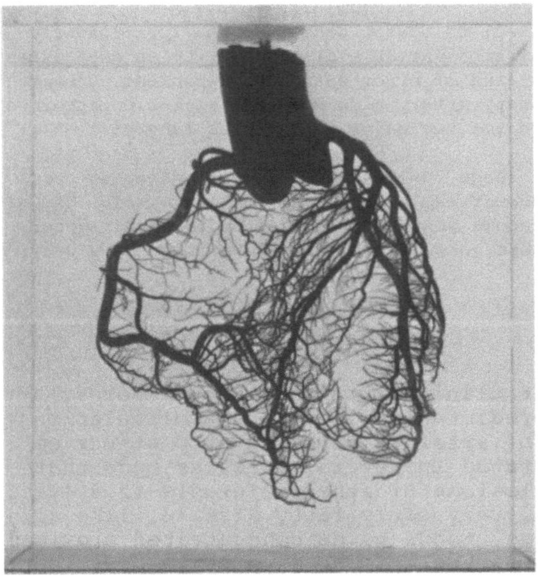

Fig. 1. Resin cast of human coronary arteries

Measurements were made on the casts with calipers, vernier gauges, and protractors. The diameter of each major vessel (or both diameters where it was elliptical) was measured from its ostium at 1 cm intervals as far as possible downstream and the diameter of each branch larger than 0.5 mm measured at its origin. Vessels smaller than 0.5 mm were treated as if they were 0.5 mm. The angle made by the left main (LMCA) and right (RCA) coronary arteries with the axial and cross sectional planes of the ascending aorta, and by the LAD and LCX arteries with each other and with the LMCA, were measured with protractors.

Results

Aorta and LMCA

The basic geometry of the proximal ascending aorta and the left main coronary artery are illustrated in Fig. 2. Immediately downstream of the cusps of the aortic valve are three roughly hemispherical outpouchings of the aortic wall, almost identical in size, known as the sinuses of Valsalva. The LMCA takes origin from a rapidly tapering, bell-shaped ostium situated near the downstream end of the left sinus of Valsalva. The angle between the LMCA and the longitudinal axis of the aorta varies considerably, but in all but one case formed an acute angle (i.e. back towards the heart), with a mean of 46 degrees.

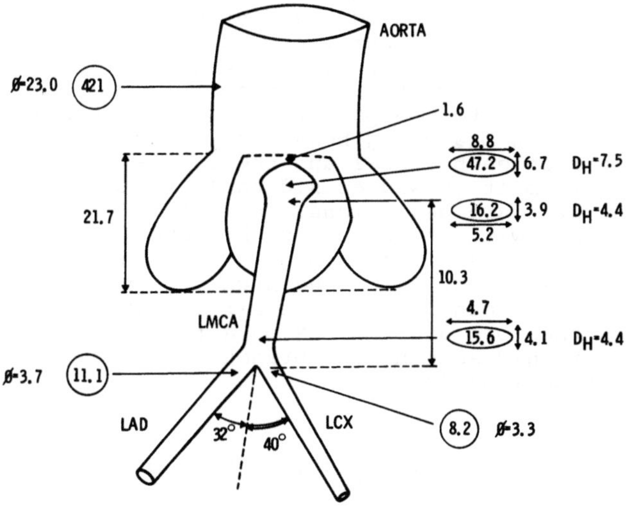

Fig. 2. Scale drawing of the ascending aorta and the LMCA, which arises at the downstream end of the sinus of Valsalva and angles backwards on to the surface of the left ventricle before giving rise to LAD and LCX. Measurements are mean values, in mm. Ringed values are areas, in mm^2. D_H: hydraulic diameter (mm)

Mean dimensions for the aorta, sinus of Valsalva, and LMCA are given in Fig. 2. There was some variation in these dimensions from specimen to specimen; coefficients of variation (SD as % of mean) were 10-20%. Linear regression analysis showed that each individual dimension correlated well with aortic diameter (correlation coefficients 0.6-0.84). In other words, the size of the sinus and LMCA were scaled to the size of the aorta and probably therefore to the body size of the subjects, which is known to determine aortic size. The one exception to this was the length of the LMCA, which varied widely (3.0-20.0 mm) and showed no correlation with any other local dimension. The LMCA was straight, and throughout its length maintained an elliptical shape. It terminated into the LAD and LCX branches. Further details of dimensions are given in Table 1.

Table 1. Dimensions of major coronary arteries. Values are means ± 1SD

	LMCA	LAD	LCX	RCA
Diameter (mm)	*4.4 ± 0.5	3.7 ± 0.5	3.2 ± 0.5	3.6 ± 0.5
Area (mm^2)	16.2 ± 3.8	11.1 ± 2.8	8.2 ± 2.5	10.2 ± 2.8
Length (mm)	10.3 ± 4.2	-	-	-
Area taper (% per cm)	4.8 ± 9.7	9.8 ± 3.3	10.5 ± 6.5	6.8 ± 2.3

* Hydraulic diameter

LAD and LCX

The angle between these two branches at the bifurcation of the LMCA ranged from 50-95 degrees with a mean value of 72 degrees. As can be seen from Fig. 2, this bifurcation was nearly symmetric relative to the axis of LMCA. Both the daughter vessels were circular throughout their lengths; their dimensions at their origins are given in Table 1. In all but two specimens, the LAD was larger than the LCX. The mean ratio of combined cross sectional areas of the two daughters to that of the parent LMCA was 1.23. The LAD followed the contour of the left ventricular surface with a single, gentle curvature throughout its length. The LCX showed much more variable curvature, with two components, following the wall of the left ventricle but at the same time curving backwards from the anterior to the posterior surface of the ventricle. Detailed information on branching behaviour of both vessels is given in Table 2. The taper of each vessel and the cross-sectional area of the branches issuing from it showed considerable variation, as can be confirmed from the standard deviations. The rate of taper appeared to correlate positively with the number of branches and negatively with the overall length of the parent vessel. The branching patterns of the LAD and LCX showed a striking and highly significant ($P < .001$) difference in the first centimetre of their length (Table 2), with the LAD having less than one and the LCX more than two branches on average. Downstream of this segment the branching behaviour of the two vessels was indistinguishable.

RCA

This vessel originated in the right aortic sinus, from an ostium shaped and located almost exactly like that of the LMCA. This bell-shaped ostium tapered rapidly over a length of 2-4 mm, and at its downstream end the RCA was circular in cross-section. In over half the specimens, a second small (1-2 mm) vessel emerged separately from this ostial region; this is an anatomical variation known as the 'conus' branch. In its initial course, the RCA ran almost at right angles to the aorta (mean 88° to long axis) with a slight rightwards angulation in the horizontal plane. After 1-2 cm it began a long, continuous curvature around to the back of the heart, following the groove between the right atrium and ventricle. This curvature (which was in a plane of approximately 135 to the long axis of the aorta), was maintained through approximately 180, with a radius of curvature of 2 cm. At the posterior end of the curve, the RCA divided up into a series of small branches supplying the back of the ventricle. The branching behaviour of the vessel during the major part of its course is given in Table 2. The vessel was in general longer than either of the left coronary arteries, with similar frequency of branching, and there were often local tortuosities in its course.

Discussion

The method of preparation of these casts was intended to maintain the vessels at physiological dimensions during setting of the resin. This appears to have been successful; mean diameters of the proximal vessels in this study correspond extremely closely to those measured from selective coronary angiograms in patients with normal coronary arteries [8,9]. The geometry of these specimens also appears typical. Anatomical studies of features such as the area ratio of the LMCA and its daughter branches [10], and the location of the ostia in the sinuses of Valsalva, the length of LMCA, and the size of the sinuses [11] gives values close to those reported here.

Table 2. Branching characteristics of major coronary arteries

Vessel segment (cm)		0–1	1–2	2–3	3–4	4–5	5–6	6–7
	LAD	0.88 ± 0.7	1.82 ± 1.1	1.41 ± 0.9	1.25 ± 0.9	1.38 ± 0.9	1.50 ± 1.0	1.27 ± 1.0
A	LCX	2.19 ± 1.1	1.87 ± 0.7	1.33 ± 1.0	1.15 ± 0.7	1.90 ± 1.2	1.86 ± 1.3	1.80 ± 0.8
	RCA	1.0 ± 1.1	1.63 ± 0.9	1.16 ± 0.8	1.5 ± 1.0	1.47 ± 0.9	1.31 ± 0.7	1.38 ± 0.9
	LAD	1.1 ± 1.63	3.1 ± 2.85	2.8 ± 2.72	1.6 ± 2.08	1.7 ± 1.81	1.8 ± 2.53	1.2 ± 1.16
B	LCX	1.2 ± 1.46	1.3 ± 1.04	0.8 ± 0.79	2.1 ± 1.52	1.7 ± 1.32	1.3 ± 1.13	1.4 ± 1.53
	RCA	1.1 ± 1.57	3.4 ± 2.79	2.8 ± 2.67	1.5 ± 2.03	1.65 ± 1.83	1.9 ± 2.48	1.0 ± 1.24
	LAD	11.1 ± 2.81	10.3 ± 2.11	8.8 ± 2.88	7.2 ± 2.75	6.2 ± 2.18	5.4 ± 1.93	4.6 ± 1.62
C	LCX	8.2 ± 2.54	7.7 ± 2.71	6.6 ± 2.44	6.0 ± 2.43	5.7 ± 2.44	5.3 ± 2.87	4.7 ± 2.71
	RCA	10.2 ± 2.75	9.4 ± 2.31	9.0 ± 2.33	8.1 ± 2.58	7.5 ± 2.46	7.2 ± 2.96	7.3 ± 2.5

A : no. of branches per segment
B : area of branches in each segment (mm^2)
C : area of parent vessel at entry to each segment (mm^2)
All values are means, ± 1SD.

In order to calculate relevant fluid mechanic parameters for these vessels, values must be assumed for blood flow velocity in the major coronary arteries, since these have not been reported in man. However, many relevant measurements of coronary blood flow in man are available. Thus, total flow is in the region of 250 ml/min [12,13] at rest, and is somewhat higher in the left than the right coronary artery [14,15, 16] . It therefore seems reasonable to assign flow rates of 150 and 100 ml/min to the left and right coronary arteries respectively, and this is supported by direct measurements which gave a mean figure for eight patients of 116 ml/min in the right coronary artery [17]. Such a division of flow is also in proportion to the relative areas of the two vessels (Table 1). Table 3 shows mean values for major fluid mechanic parameters for the vessels in this study, assuming the above flow rates, a heart rate of 1.2 Hz, and flow apportioned to LAD and LCX in proportion to cross-sectional area. Reynolds number is based on mean velocity and vessel diameter, assuming a kinematic viscosity of 0.04 cm^2/sec for blood. Entrance length and shear stress are calculated for steady Poiseuille flow in a straight pipe.

Table 3. Inlet fluid mechanic characteristics of major human coronary arteries. Values are means ± 1SD

	LMCA	LAD	LCX	RCA
Mean resting flow velocity (cm/sec)	16.0 ± 3	13.6 ± 3	13.8 ± 3	17.8 ± 6
Reynolds number	171 ± 20	128 ± 28	111 ± 15	152 ± 25
Wormersley parameter	3.28 ± 0.44	2.81 ± 0.34	2.41 ± 0.39	2.68 ± 0.39
Entrance length (cm)	2.2 ± 0.3	1.4 ± 0.4	1.1 ± 0.2	1.6 ± 0
Entrance length (diameters)	5.1 ± 2.2	3.8 ± 0.8	3.3 ± 0.5	4.5 ± 0.7
Mean wall shear-stress ($dynes/cm^2$)	12.1 ± 3.2	12.2 ± 2.9	14.5 ± 5.6	17.2 ± 8.9

The values in Table 3 must be interpreted with caution. As in all major arteries, mean velocity is something of an abstraction, since flow is highly unsteady. In man peak velocities are two to three times the mean, and the instantaneous velocity during diastole is much greater than that during systole [18,19]. Furthermore, assumption of a Poiseuille profile will underestimate shear stress in these entrance regions. Thus it seems likely that shear stress at the wall in these vessels will fluctuate from zero to at least two or three times the values in Table 3 during the cardiac cycle. Furthermore, direct measurements in animals have shown that peak coronary flow velocity can increase by a factor of three or four in response to exercise and emotion [20,21,22]. Thus peak wall shear stresses in the region of 100 $dynes/cm^2$ may be expected in all these vessels during strenuous exercise.

A recent study of the location of early atheroma in the human coronary bed [3] demonstrated that disease is confined to the proximal region of all these major branches. It is therefore of interest to know how these fluid mechanic parameters change with distance downstream. In Table 4, values have been computed for each vessel at a downstream site chosen to correspond to the point beyond which no disease was seen in the previous study [3]. In calculating local flow rates, the assumption was made that division of flow between the parent vessel and its small branches is proportional to cross-sectional area.

Table 4. Downstream fluid mechanic parameters of major human coronary arteries. Values are means ± 1SD

	LAD	LCX	RCA
Distance downstream (cm)	6	5	10
Mean diameter (mm)	2.4 ± 0.43	2.5 ± 0.74	2.8 ± 0.55
Mean resting flow velocity (cm/sec)	28 ± 15	12 ± 7	19 ± 13
Reynolds number	153 ± 64	70 ± 34	120 ± 57
Wormersley parameter	1.5 ± 0.27	1.7 ± 0.47	1.8 ± 0.37
Mean wall shear stress (dynes/cm^2)	42 ± 29	18.4 ± 16	25.7 ± 22

Because of anatomic variation, downstream velocities calculated in this way show wide scatter, which emerges in the figures for Reynolds number and shear stress. However, the differences between upstream and downstream values are modest, and only reach significance in the case of shear stress in LAD and Reynolds number in LCX.

It is clear that even peak shear stresses in these vessels are well below the levels shown by Fry [23] to cause acute endothelial damage. Furthermore, the very similar values of fluid mechanic parameters obtaining at both upstream and downstream sites in LAD, LCX and RCA offer no obvious insight into the very different incidence of early disease in these vessels or its location. On the other hand, the local anatomic arrangement of the vessels may be important, and a number of observations in this study point in that direction. For example, in the immediate vicinity of the LAD-LCX bifurcation, the level of shear stress should be altered by the branching process. In this, the exact angle between these two daughter vessels is important, and the range previously indicated (50° to 95°) implies that wall shear stress may vary widely, depending on the exact geometry of the individual subject. A further anatomic feature of interest is the difference in branching patterns in the first centimetre (2-4 diameters) of the LAD and LCX (Table 2). It should be noted that the relative paucity of branches in the LAD corresponds to a region of relative sparing from early disease [3]. Finally, there is the long, continuous curvature of the RCA. The Dean number ($D = 4\ Re\ (2\delta)^{0,5}$, where Re is Reynolds number and δ = vessel radius over radius of curvature) associated with flow in this curvature ranges between approximately 250 (rest) and 1700 (exercise). For such values there will be a significant alteration of the velocity profile and wall shear stress due to curvature. This may be of relevance to the tendency for early atheroma to appear on the inner wall of this curvature [3].

Acknowledgements. This work was supported by National Science Foundation grant CME-80-01701, and by a Travel Grant from the Wellcome Trust. We wish to thank Chris Lee for technical assistance with preparation of the casts.

References

1. Mitchell JRA, Schwartz CJ (1965) Arterial disease. Blackwell, Oxford
2. Kjaernes M, Svindland A, Walle L, Willoe SO (1981) Localization of early arteriosclerotic lesions in an arterial bifurcation in humans. Acta Path Microbiol Scand Sect A 89: 35-40
3. Fox B, James K, Morgan B, Seed A (1982) Distribution of fatty and fibrous plaques in young human coronary arteries. Atherosclerosis 41: 337-347
4. Giddens DP, Zarins CK, Glagov S, Bharadraj BK, Ku DN Flow and atherogenesis in the human carotid bifurcation. This volume, p
5. Svindland AD, Walloe L Localization of early atherosclerotic lesions in carotid and coronary bifurcations in humans. This volume, p
6. Caro CG, Fitzgerald JM, Schroter RC (1971) Atheroma and arterial wall-shear. Proc R Soc Lond B 177: 109-159
7. Fry DL (1972) Localizing factors in arteriosclerosis. In: Likeoff W, Segal BL, Insull W, Moyer JH (eds) Atherosclerosis and coronary heart disease. Grune & Stratton, New York
8. Viewig WVR, Alpert JS, Hagan AD (1976) Caliber and distribution of normal coronary arterial anatomy. Cathet Cardiovasc Diagn 2: 269-280
9. MacAlpin RN, Abbasi AS, Grollman JH, Eber L (1973) Human coronary artery size during life. Radiology 108: 567-576
10. Hutchins GM, Miner MM, Boitnott JK (1976) Vessel caliber and branch-angle of human coronary artery branch-points. Circ Res 38: 572-576
11. McAlpine WA (1975) Heart and coronary arteries: an anatomical atlas. Springer, Berlin Heidelberg New York
12. Cohen A, Gallagher JP, Luebs ED, Varga Z, Yamanaka J, Zaleski EJ, Bluemchen G, Bing RJ (1965) The quantitative determination of coronary flow with a positron emitter (rubidium-84). Circulation 32: 636-649
13. Knoebel SB, McHenry PL, Stein L, Sonel A (1967) Myocardial blood flow in man as measured by a coincidence counting system and a single bolus of 84 RbCl. Circ 36: 187-196
14. Ross RS, Ueda K, Lichtlen PR, Rees JR (1964) Measurement of myocardial blood flow in animals and man by selective injection of radioactive inert gas into the coronary arteries. Circ Res 15: 28-41
15. Pitt A, Friesinger GC, Ross RS (1969) Measurement of blood flow in the right and left coronary artery beds in humans and dogs using the ^{133}Xenon technique. Cardiovasc Res 3: 100-106
16. Horwitz LD, Curry GC, Parkey RW, Bonte FJ (1974) Differentiation of physiologically significant coronary artery lesions by coronary blood flow measurements during isoproterenol infusion. Circulation 49: 55-62
17. Folts JD, Rowe GG, Kahn DR, Young WP (1979) Phasic changes in human right coronary blood flow before and after repair of aortic insufficiency. Am Heart J 97: 211-215
18. Cole JS, Hartley CJ (1977) The pulsed Doppler coronary artery catheter. Circulation 56: 18-25
19. Marcus M, Wright C, Doty D, Eastham C, Laughlin D, Krumm P, Fastenow C, Brody M (1981) Measurements of coronary velocity and reactive hyperaemia in the coronary circulation in humans. Circ Res 49: 877-891
20. Vatner SF, Higgins CB, Franklin D, Braunwald E (1972) Role of tachycardia in mediating the coronary hemodynamic response to severe exercise. J Appl Physiol 32: 380-385
21. Van Citters RL, Franklin DL (1969) Cardiovascular performance of Alaska sled dogs during exercise. Circ Res 24: 33-42
22. Vatner SF, Franklin D, Higgins CB, Patrick T, White S, Van Citters RL (1971) Coronary dynamics in unrestrained conscious baboons. Am J Physiol 221: 1396-1401
23. Fry DL (1968) Acute vascular endothelial changes associated with increased blood velocity gradients. Circ Res 22: 165-197

Discussion

CARO: Just one comment and this is with respect to smooth muscle tone. It may make no difference, but one's understanding is that vessels lose tone after death. I would add, and understand that I am a complete amateur here, that you can achieve some tone in such arteries with the use of noradrenaline.

SEED: You can, but that is over a time scale which is much shorter; many of these arteries have been frozen prior to our actually perfusing them. I think for about 48 hours you can get some life out of the smooth muscle.

CARO: Would it be interesting to compare your data with results obtained from arteriograms?

SEED: First, these have no smooth muscle tone at all. Next, when we do what you suggest, given the scatter that you get in the angiographic literature, we are right in the middle of the range both on diameters and cross-sectional areas for the major vessels. I think smooth muscle tone is here, but I do not think it is making an important contribution to dimensions except if there is a spasm or something of that sort.

SCHMID-SCHÖNBEIN: Your localization, central localization of the disease, does that apply to older patients as well, to an older populace.

SEED: We have only looked at fatty and fibrous plaques. We have not looked at disease which is advanced enough to be complicated, but certainly the graphs I showed you of fatty disease are almost superimposable on these for fibrous disease. In other words, we feel very strongly that we are looking at something which progresses and which is defining the localization of the more mature disease.

Flow Patterns in Model and Natural Branching Vessels

T. Karino, M. Motomiya, and H. L. Goldsmith

Introduction

Clinical and postmortem studies indicate that both primary atherosclerotic lesions and the deposition of platelet thrombi occur preferentially at the entrances of branching arteries and curved segments where flow is likely to be disturbed and separation of streamlines from the vessel wall and formation of vortices may occur (Mustard et al. 1962; Roach 1977).

Hence, to elucidate the possible connection between blood flow and the localized genesis and development of atherosclerosis and thrombosis, a considerable amount of work, both theoretical and experimental, has been carried out by many investigators including ourselves (Goldsmith and Karino 1978).

In our previous microrheological investigation on the flow behavior and interaction of red cells and platelets in an annular vortex formed downstream of sudden tubular expansions (serving as models of axisymmetric stenoses), we demonstrated that the vortex provided favorable conditions for the genesis, growth and trapping of platelet aggregates as well as for the enhanced adhesion of cells to the vessel wall in the region of disturbed flow (Karino and Goldsmith 1977, 1979a,b).

In the platelet adhesion experiments, where the effect of local flow patterns on the initial deposition of platelets onto the vessel wall was investigated in detail, it was convincingly shown that both in steady and pulsatile flow, platelet adhesion was localized within the vortex and downstream on either side of the reattachment point with a local minimum at the reattachment point itself. The localization of cell adhesion became more pronounced as the hematocrit increased. Surprisingly though, the adhesion peak in the vortex decreased and flattened out as the Reynolds number increased. These results were inconsistent with a diffusion-controlled platelet adhesion which predicts an increase in adhesion number density with increasing shear rate.

The mechanism of this localization of platelet adhesion was explained as being due to the enhanced transport of platelets to the wall along the radially directed streamlines existing in the vicinity of the reattachment (stagnation) point as shown in Fig. 1 (Karino and Goldsmith 1979b). The significance of the above finding is that this same mechanism may apply not only to cellular elements but also to macromolecules such as plasma proteins and lipids including cholesterol. The effect will be further enhanced by the convective motion of red cells around the reattachment point. This will, in turn, result in an increased diffusion and convective transport of materials in the blood as well as across the endothelium, eventually leading to their accumulation in the vessel wall in the regions of disturbed flow on either side of the reattachment (stagnation) point. It follows from this hypothesis that the deposition of platelet thrombi and early atheroscler-

Fluid Dynamics as a Localizing Factor for Atherosclerosis
Editors: G. Schettler et al.

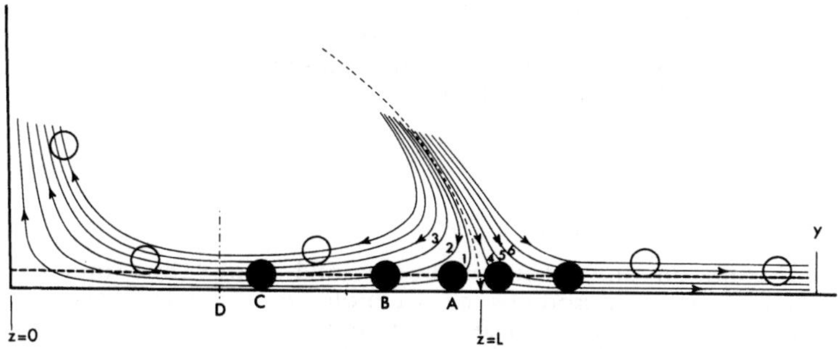

Fig. 1. Schematic diagram of flow patterns around a reattachment (stagnation) point showing convective transport of particles from the bulk flow toward the vessel wall along the curved streamlines having a pronounced radial velocity component. The *black spheres* are carried by the flow along the radially directed streamlines and approach to within the critical distance of the wall, indicated by the *thick dashed line* (one particle radius from the surface), and collide with the vessel wall at points A to C in the vortex region, and at corresponding points downstream of the reattachment point. The *open circles* are spheres which do not come close enough to collide with the wall (from Karino and Goldsmith 1979b)

tic lesions will be localized wherever there is a stagnation point (or a reattachment point if it is associated with flow separation) where blood cells are carried by the flow toward the vessel wall along curved streamlines having a pronounced radial velocity component. Thus, in the circulation, one would predict a relatively higher risk of thrombosis and atherosclerosis not only in regions of separated flow (due to enhanced platelet adhesion and atherosclerotic changes on either side of the reattachment point) such as downstream of the aortic valve, mural thrombi and stenoses, but also in all the branching arteries at the flow divider where there is a stagnation point.

The above hypothesis differs from that of a shear rate-dependent atherogenesis proposed by Fry (1968) and Caro et al. (1971) and may provide an alternate explanation for the localized genesis and development of atherosclerosis and thrombosis in the circulation. Thus, it became important to obtain exact and precise flow patterns and distribution of shear rate (or shear stress) existing in various branching arteries and to correlate these results with the clinical findings of atherosclerosis and thrombosis.

In this paper, we summarize our latest knowledge of the detailed flow patterns and distributions of fluid velocity and shear rate obtained in various glass models of T-junctions, as well as in isolated transparent natural branching arteries prepared from dogs and humans post-mortem, by a method recently developed in our laboratory (Karino and Motomiya 1982).

Methods

Glass model T-junctions having various branching angles and tube diameter ratios were constructed either by fitting and gluing together two pieces of thin-walled glass tubing at desired angles, or by glass blowing. Transparent natural branching arteries were prepared from dogs and humans post-mortem, by fixing the vessels under the physiological transmural pressures, dehydrating with ethanol and rendering transparent using conventional tissue clearing techniques. Flow was visualized by taking 16 mm cine films of the motions of tracer polystyrene microspheres in dilute suspensions undergoing steady flow at various flow rates. The developed films were subsequently projected on a drafting table and analyzed frame by frame.

Results

I. Flow Patterns in Glass Model T-Junctions

A series of investigations on the flow patterns and distributions of fluid velocity and shear rate in models of branching vessels were carried out using thin-walled 3 mm diameter glass T-junctions having branching angles of 90°, 60°, 45° and 30°, and diameter ratios d/D (side to main tube) from 0.33 to 1.0. Experiments were carried out over a range of inflow Reynolds numbers, Re_o, from 10 to 350 at various flow rate ratios Q_1/Q_o (main branch to parent tube).

a) General Flow Patterns in 90° Uniform Diameter T-Junctions. When flow entered the main tube, formation of paired spiral secondary flows and recirculation zones symmetrical about the common median plane of parent and daughter tubes were demonstrated over a wide range of Re_o and Q_1/Q_o. Figure 2 shows an example of flow patterns (traced from the paths of tracer particles) in a 3 mm uniform diameter, 90° square T-junction when viewed normal to the common median plane. Under these flow conditions (Q_1/Q_o = 0.2), a large recirculation zone was formed in the main tube at the entry of the daughter tube, and a small secondary corner recirculation zone at the entry of the side branch. The formation and the sizes of these recirculation zones were largely dependent on the flow ratio Q_1/Q_o as well as on Re_o. When observed along the common median plane, it was found that the recirculation zones in both main and side tubes consisted of a pair of spiral secondary flows located symmetrically on both sides of the common median plane. Particles en-

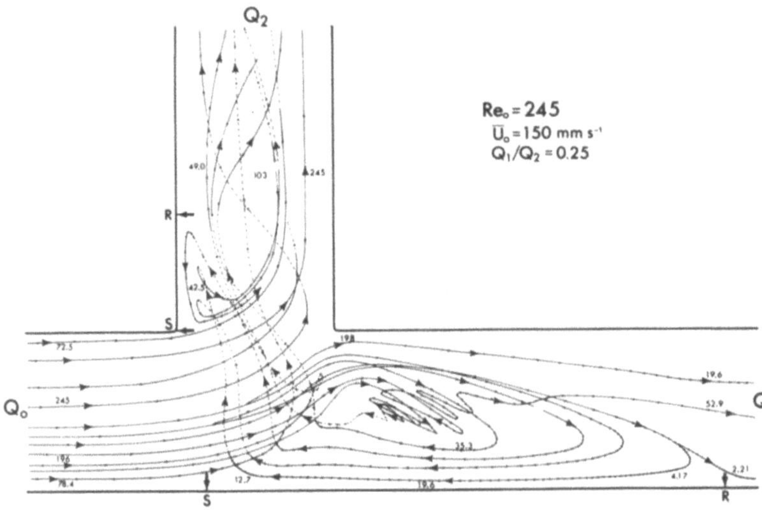

Fig. 2. Flow patterns in the common median plane of a uniform 3 mm diameter 90° T-junction. The *arrows* at S and R indicate the respective separation and reattachment points of the flow and the numbers on the streamlines indicate the particle linear velocities in mm s^{-1} (from Karino et al. 1979c)

tering the main recirculation zone described complicated orbits, some of them rejoining the flow through the main tube, others crossing above or below the mainstream in the junction region far out of the median plane (indicated by the dashed path lines in Fig. 2) to the side tube and side recirculation zone downstream of which a counter-rotating double helicoidal flow developed. When the degree of occlusion of the main tube was gradually reduced, the large recirculation zone became smaller and eventually disappeared as the flow ratio was reversed ($Q_1/Q_0 = 0.8$), while the recirculation zone in the side branch grew in size. When flow entered the side tube, paired connected recirculation zones formed only when one daughter tube was much occluded.

b) Effect of Branching Angle θ and Tube Diameter Ratio d/D. Figure 3 shows the measured critical Reynolds number, Re_c, for the onset of the main and side recirculation zone as a function of relative flow rate Q_1/Q_0 (main branch/parent tube). As evident from Fig. 3 (left), Re_c for the formation of the main recirculation zone was the lowest at 90° for all Q_1/Q_0, whereas for the side recirculation zone it decreased as the branching angle increased from 45° to 135°. With decreasing tube diameter ratio d/D (or the area ratio α), Re_c for both the main and side

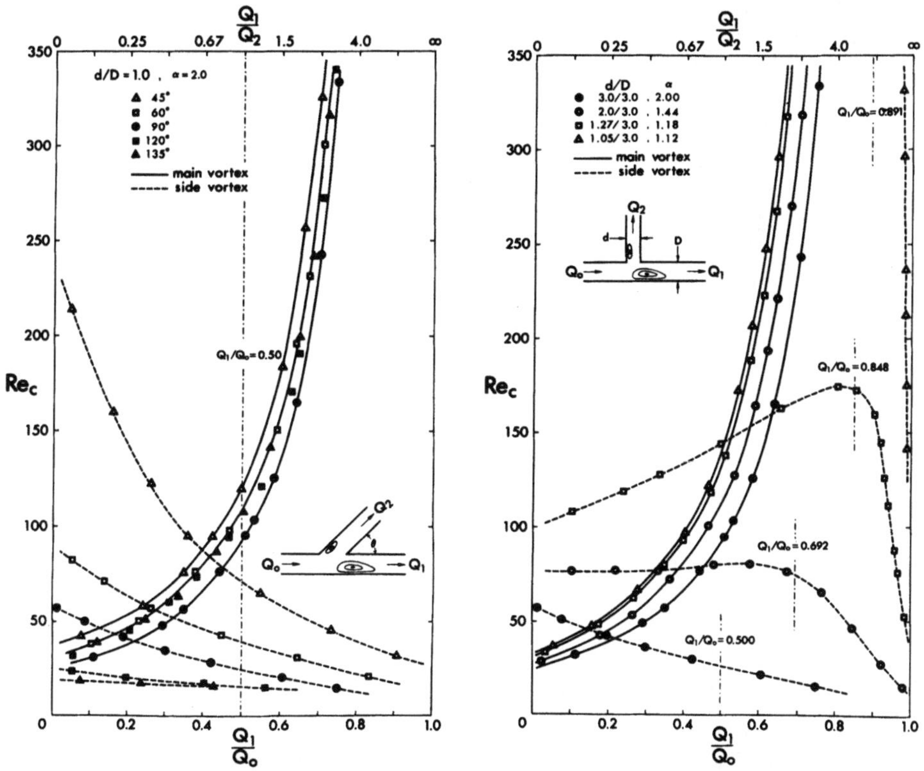

Fig. 3. Plot of measured critical Reynolds number, Re_c, for the formation of main and side recirculation zones against relative flow rate Q_1/Q_0 or Q_1/Q_2. Left: Effect of branching angles on the formation of main (*solid lines*) and side (*dashed lines*) recirculation zones in uniform diameter T-junctions. Right: Effect of tube diameter ratio on the formation of main and side recirculation zones in various 90° T-junctions. The vertical broken dashed lines indicate geometrical flow rate ratio for each T-junction (from Karino and Goldsmith 1980)

recirculation zone drastically increased as shown in Fig. 3 (right). For T-junctions having a side tube of 3 and 2 mm diameter (d/D = 1 and 2/3), both the main and side recirculation zone were formed at geometrical flow ratios (calculated by assuming that flow in the two branches is distributed proportionally to their cross-sectional area). With decreasing side tube diameter, the main recirculation zone which consisted of a pair of spiral secondary flows located symmetrically on both sides of the common median plane of the T-junction, became smaller and was confined to a thin layer adjacent to the tube wall, wrapped around the predominant main flow. In T-junctions with d/D < 1/2, no main recirculation zone was formed at the geometrical flow ratio. Thus, for d/D = 1.27/3.0 only the side recirculation zone and in T-junctions with the smallest side tube (d/D = 1.05/3.0), neither main nor side recirculation zone was formed. When one of the branches was completely occluded, another type of large recirculation zone which occupied the entire lumen was formed in the dead-end branch.

c) Effect of Radius of Curvature of the Walls. This was studied by comparing Re_c and the size of the recirculation zone in the rounded (radii of curvature ~ d/2) and square (radii of curvature < 0.02D) T-junctions. As shown in Fig. 4, there were significant differences between the two types of T-junctions. Thus, at a given Re_o and Q_1/Q_o, a much larger main recirculation zone with more rapidly circulating fluid existed in the rounded than in the square T-junction. These date show that formation of the side recirculation zone is largely affected by curvature of the wall at the bend opposite the flow divider, while the formation of the main recirculation zone is affected by the curvature at the flow divider.

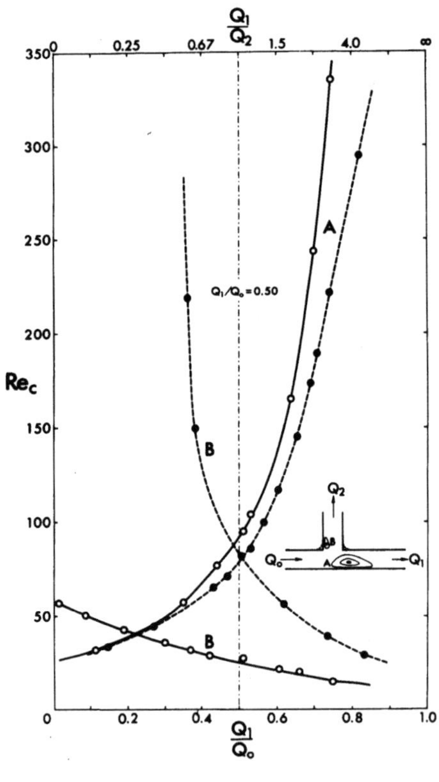

Fig. 4. Plot of the measured Re_c for the formation of main (A) and side (B) recirculation zone against Q_1/Q_o as in Fig. 3. The figure shows the effect of radius of curvature at the junction of the formation of main and side recirculation zones in 90°, 3 mm uniform diameter square (*solid lines*) and rounded (*dashed lines*) T-junctions (from Karino and Goldsmith 1980)

II. Flow Patterns in Isolated Transparent Natural Branching Arteries

To relate the results obtained in glass model T-junctions to the flow in the actual circulation, transparent segments of a dog abdominal aorta containing branches of the celiac, cranial-mesenteric and right and left renal arteries, and a human carotid artery bifurcation were prepared. The flow patterns were studied in detail using the same techniques as in the glass model T-junctions.

a) Flow patterns in Dog Aortic T-Junctions. The general flow patterns at each T-junction were similar to those observed in glass models, as illustrated in Fig. 5. The figure shows the formation of paired spiral secondary flows at the aorto-celiac artery junction (d/D = 4.0 mm/7.2 mm) at the geometrical flow rate ratio. Critical Reynolds numbers, Re_c, for the formation of the spiral secondary flow at each junction were also measured as a function of the flow rate ratio in the two daughter tubes. The results for the aorto-celiac artery junction showed that at the geometrical flow rate ratio, $Q_1/Q_0 = 0.72$, spiral secondary flows were formed at $Re_0 \sim 300$, grew bigger with increasing Re_0, and reached a fully disturbed flow at $Re_0 \sim 900$ (the largest orbit then reached the opposite wall and the paired spiral secondary flow completely encircled the undisturbed main flow). With further increasing Re_0, the recirculation zone grew in length and then, at $Re_0 \sim 1200$, there was a sudden transition from disturbed to turbulent flow. An interesting observation was made on the geometrical structure of the natural vessel. Close examination of the wall at each function revealed that curvature of the wall at the flow divider was very sharp (< 90°) with a sudden decrease in aortic diameter distal to the branch, whereas that at the opposite corner was gently rounded. From the results obtained in the glass model T-junctions described earlier, this represents the optimum condition for minimizing the size of both main and side recirculation zone which may form in the aortic T-junction. In fact, it was found that under physiological conditions, $Re_0 \sim 700$, and at the geometrical flow ratio, there was no side recirculation zone, and the main

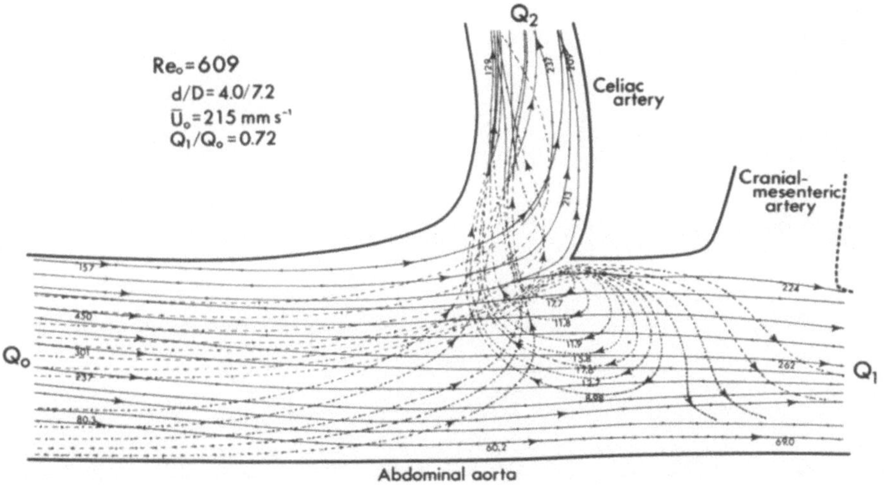

Fig. 5. Detailed flow patterns at the aorto-celiac artery junction of a dog abdominal aorta, showing the formation of paired, thin-layered spiral secondary flows on both sides of the common median plane, adjacent the vessel wall wrapping the undisturbed mainstream. The numbers on the streamlines indicate the particle linear velocities in mm s^{-1} (from Karino et al. 1982)

recirculation zone was small and confined to a thin layer close to the wall surrounding the mainstream. Hence, in the circulation, conditions at arterial branches appear to be optimal for minimizing the formation of recirculation zones.

b) Flow Patterns in the Human Carotid Artery Bifurcation. It was found that paired spiral secondary flows, located symmetrically on both sides of the common median plane, were formed in the carotid sinus. This occurred over a wide range of inflow Reynolds numbers, Re_o, based on the common carotid artery, and flow rate ratios, Q_1/Q_o (internal/common), as shown by a plot of critical Reynolds number for the formation of spiral secondary flow, Re_c vs. Q_1/Q_o in Fig. 6. At the geometrical flow rate ratio $Q_1/Q_o = 0.65$ (indicated by the vertical dashed line in Fig. 6), a recirculation zone was formed only in the internal carotid artery at $Re \sim 170$, grew in size with increasing Re_o, but was always confined within the carotid sinus. The recirculation zone increased in length from about 4 mm at $Re_o = 300$ to a maximum of 9 mm at $Re_o > 800$. Figure 7 gives an example of detailed flow patterns (upper) and distributions of fluid velocity and wall shear rate (lower) in the carotid sinus. As shown in the upper figure, particles were deflected at the flow divider, traveled laterally and very slowly along the vessel wall above and below common median plane, almost at right angles too, and encircling the mainstream. They then changed direction, moving back along the outer wall of the internal carotid ar-

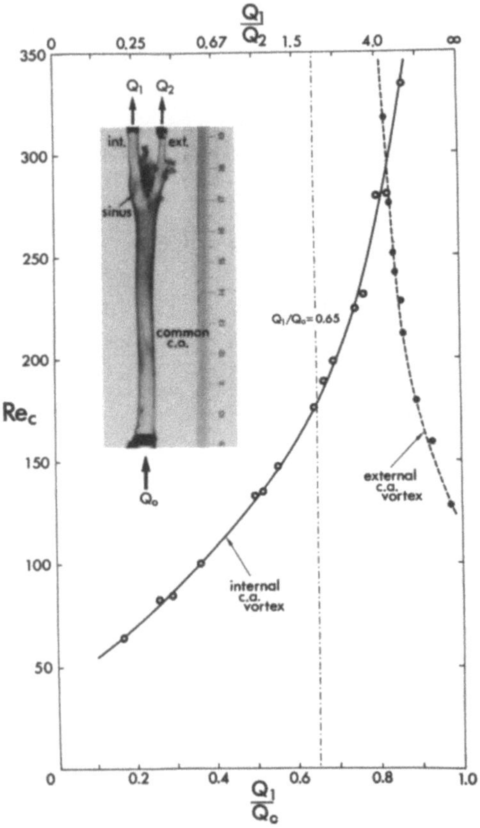

Fig. 6. Plot of measured Re_c for the formation of recirculation zones in a human carotid artery bifurcation against the flow rate ratio, Q_1/Q_o (from Motomiya and Karino 1982)

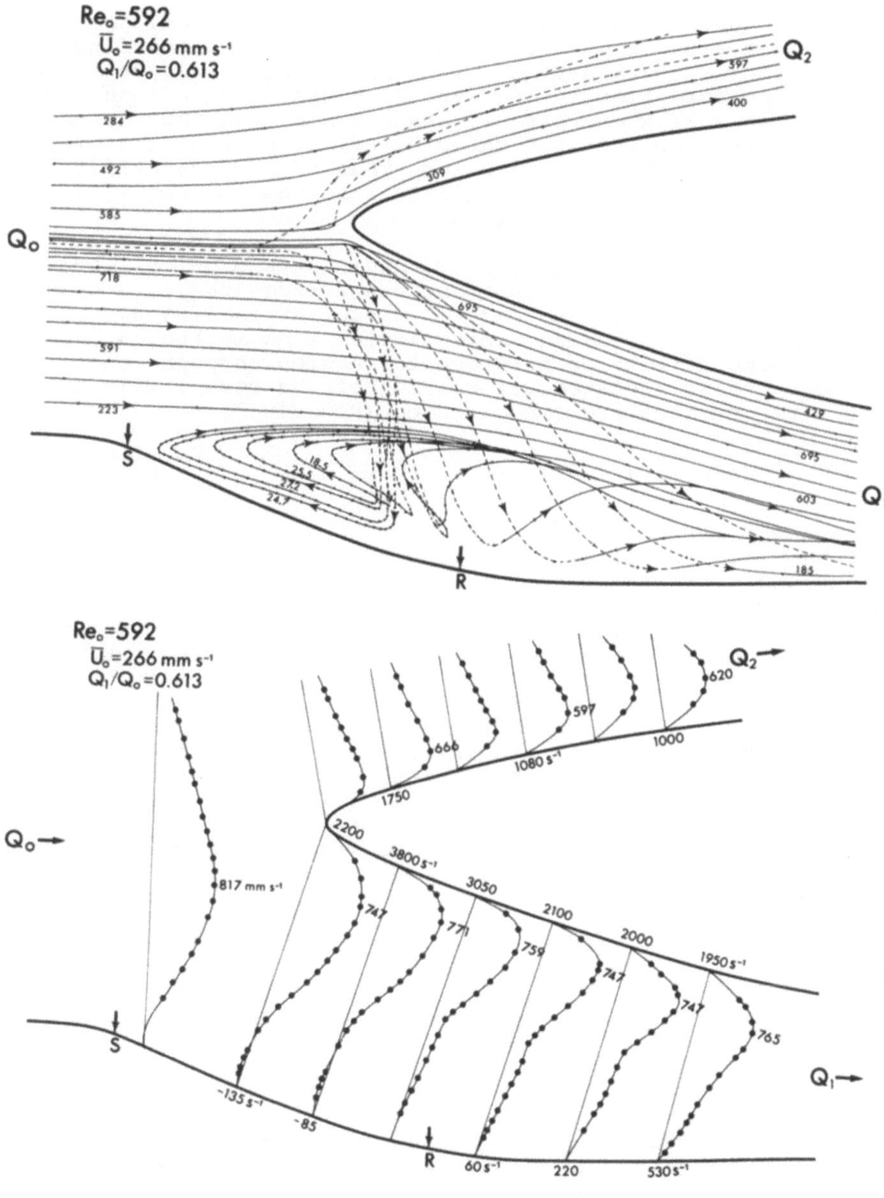

Fig. 7. Detailed flow patterns (*upper*) and the distribution of fluid velocity and wall shear rate on the common median plane of the parent and daughter vessels (*lower*) in a human carotid artery bifurcation at the geometrical flow rate ratio. The arrows at S and R denote the respective separation and stagnation points. The numbers on the streamlines indicate the particle linear velocities in mm s^{-1} (from Motomiya and Karino 1982)

tery at the site of the sinus, describing spiral orbits in a recirculation zone before rejoining the mainstream. Downstream of the stagnation point (indicated by R in the figure), a strong counter-rotating double helicoidal flow developed. The velocity profiles (lower figure) were extremely skewed towards the inner wall of the bifurcation, creating a high shear field along the vessel walls downstream of the flow divider where there was a stagnation point. Due to the presence of the paired spiral secondary flow in the carotid sinus, the wall shear rate and hence the wall shear stress, changes sign and becomes negative at the separation point (indicated by S); it becomes positive again downstream of the stagnation point. Thus, in the carotid sinus, there are regions where the vessel wall is stretched or compressed by the counter-directed wall shear stresses.

The above results strongly suggest that, under physiological conditions ($Re_o \sim 600$, $Q_1/Q_o \sim 0.7$ [Kristiansen and Krog 1962]), a paired recirculation zone exists in the carotid sinus. It is also speculated that the formation of a standing recirculation zone in the carotid sinus may have some effects on baroreceptor action through counter-directed, shear stress-mediated deformation of the vessel wall in the sinus where nerve endings are concentrated (Rees 1967).

Conclusion

The use of a newly developed technique for making blood vessels transparent has, for the first time, enabled us to obtain the exact flow patterns and distributions of fluid velocity and shear rate existing in various regions of disturbed flow in the circulation. The results demonstrated convincingly that flow separation and formation of paired spiral secondary flows and recirculation zones previously observed in glass model T-junctions must occur in vivo at aortic T-junctions and at the carotid bifurcation in man. Such flows will affect local mass transfer and the interactions of blood cells, especially the platelets, with the vessel wall and in turn may lead to the localized occurrence of thrombosis and atherosclerosis in these regions.

Acknowledgements. This work was supported by MRC, Canada (MT-7084, MT-1835) and the Quebec Heart Foundation. The authors gratefully acknowledge the technical assistance of C. Artigas and E. De Heuvel.

References

Caro CG, Fitz-Gerald JM, Schroter RC (1971) Arterial wall shear. Observation, correlation and proposal of shear dependent mass transfer mechanisms for atherogenesis. Proc R Soc Lond B 177: 109-159

Fry DL (1968) Acute vascular endothelial changes associated with increased blood velocity gradients. Circ Res 22: 165-197

Goldsmith HL, Karino T (1978) Mechanically induced thromboemboli. In: Hwang NHC, Gross DR, Patel DJ (eds) Quantitative cardiovascular studies - clinical and research applications of engineering principles. University Park Press, Baltimore, pp 289-351

Karino T, Goldsmith HL (1977) Flow behaviour of blood cells and rigid spheres in an annular vortex. Phil Trans R Soc Lond B 279: 413-445

Karino T, Goldsmith HL (1979a) Aggregation of human platelets in an annular vortex distal to a tubular expansion. Microvasc Res 17: 217-237

Karino T, Goldsmith HL (1979b) Adhesion of human platelets to collagen on the walls distal to a tubular expansion. Microvasc Res 17: 238-262

Karino T, Kwong HHM, Goldsmith HL (1979c) Particle flow behavior in models of branching vessels. I. Vortices in 90° T-junctions. Biorheology 16: 231-248

Karino T, Goldsmith HL (1980) Disturbed flow in models of branching vessels. Trans Am Soc Artif Intern Organs XXVI: 500-506

Karino T, Motomiya M Flow through a venous valve and its implication in thrombogenesis. Submitted to Circ Res
Karino T, Motomiya M, Goldsmith HL Flow patterns in dog aortic T-junctions. Will be submitted to Circ Res
Kristiansen K, Krog J (1962) Electromagnetic studies on the blood flow through the carotid system in man. Neurology 12: 20-22
Motomiya M, Karino T Flow patterns in the human carotid artery bifurcation. Submitted to Circ Res
Mustard JF, Murphy EA, Rowsell HC, Downie HG (1962) Factors influencing thrombus formation in vivo. Am J Med 33: 621-647
Rees PM (1967) Observations on the fine structure and distribution of presumptive baroreceptor nerves at the carotid sinus. J Comp Neur 131: 517-548
Roach MR (1977) The effect of bifurcations and stenoses on arterial disease. In: Hwang NHC, Normann NA (eds) Cardiovascular flow dynamics and measurements. University Park Press, Baltimore, pp 489-593

Discussion

BORN: Did you use washed platelets for all of this?

KARINO: Most of our platelet aggregation studies were carried out using platelet-rich plasma prepared from heparinized or citrated blood, but we also did some experiments with washed platelets suspended in Tyrodes albumin solution. The rate and extent of platelet aggregation in an annular vortex were the highest in heparinized platelet-rich plasma, and they were the lowest in washed platelets.

BORN: Where aggregates form the way you describe, there is plenty of delay time and the question is how are they activated? In platelet-rich plasma the question is whether this is by shear alone or whether a chemical activator is necessary? My other comment is clinical. There is good evidence that these kinds of aggregates get swept from the common carotid and cause transient ischemic attacks or strokes of the brain. Under what conditions do these aggregates escape?

KARINO: As for the first comment, since platelet aggregates were formed only within a narrow range of shear rate, there certainly is a favorable range of shear rate for platelet aggregation in the vortex. However, there is another fact which is that such platelet aggregation in the vortex is easily suppressed or prevented by the addition of small amounts of membrane stabilizing drugs such as propranolol and chlorpromazine. This suggests that the platelet membrane property is also an important factor. As for the second comment, I have to explain the difference between the annular vortex, which forms downstream from stenoses, and various vortices and recirculation zones which form in the circulation. The recirculation zone which forms in the carotid sinus is quite different from the annular vortex. In an annular vortex, streamlines are closed and blood cells stay in the vortex for long periods of time interacting with each other and sometimes forming platelet aggregates. In contrast to this, in the carotid sinus, streamlines in the recirculation zone are not closed and there is always an inflow of new platelets from the mainstream which leave the recirculation zone after one or two spiral orbits. Therefore, it is not likely that freely floating platelet aggregates, as seen in an annular vortex, form in the carotid sinus. However, because of the presence of a recirculation zone in the carotid sinus, interactions of blood cells with the vessel wall would be enhanced on either side of the stagnation point. This will affect local mass transfer and enhance the deposition of platelets and leukocytes onto the vessel wall, and this in turn may lead to the localized occurrence of atherosclerosis and thrombosis in the carotid sinus. If the thrombi get swept away, they may plug cerebral arteries and cause transient ischemic attacks and strokes.

NAUMANN: I would like to make one comment about your work and that is that one has to be very careful about extrapolating from in vitro studies of flow on platelets and red cells to what happens in vivo. If you look at platelets ultrastructurally, they are almost invariably altered as a result of the isolation technique. This suggests that they are of augmented adhesiveness to one another, and therefore they are more likely to adhere in areas of flow disturbance. In all of the work of which that I am aware, in vivo platelets did not adhere to endothelium unless the endothelium is very severely damaged. For in vitro preparations, you are working with a thrombogenic surface; but in vivo, we have a non-thrombogenic surface.

KARINO: Yes, it is true that when we take blood from a human body we might activate some platelets, but we tested all different anti-coagulants. We also washed the platelets and tried to investigate the effect of flow for different preparations of platelets. The aggregation or adhesion which we found may or may not happen in vivo, but this is another question.

Geometric Effects on the Hemodynamic Environment of the Arterial Wall: A Basis for Geometric Risk Factors?

M. H. Friedman, O. J. Deters, F. F. Mark, C. B. Bargeron, and G. M. Hutchins

Abstract

We have proposed that some of the variability among individuals and populations in the location and rate of development of atherosclerosis is due to corresponding variations in the detailed geometry of susceptible arterial segments. The vascular geometry indirectly affects the atherogenic process by mediating the hemodynamic environment of the intima. Those geometric features which enhance atherogenic hemodynamic stresses can be regarded as "geometric risk factors". Protocols have now been implemented to identify such factors; initial results suggest that, at the aortic bifurcation, two risk factors are branch angle variations and an offset flow divider tip.

Much of the credit for the recent decline in mortality due to cardiovascular disease has been attributed to the identification of certain aspects of an individual's lifestyle, genetic makeup, or disease state which appear to predispose to atherosclerotic involvement. A recent tabulation by Anyanwu et al. (1981) lists 21 such "risk factors" for arteriosclerosis. Notwithstanding the large number of risk factors which have been identified or proposed (Hopkins and Williams 1981), they are incapable of explaining half of the variance in arterial disease among individuals and populations (McGill 1978). One conclusion that has been drawn from this situation is that there are other risk factors for atherosclerosis which have not yet been identified (Fishman et al. 1981).

In 1975, on the basis of our numerical studies of fluid dynamics in branches having differing geometries, we proposed (Friedman et al. 1975) that individual variations in arterial geometry might form the basis for what we then called "architectural risk factors". Since that time, a principal objective of our hemodynamic research has been to test this concept, which is supported by a proposed mechanism involving hemodynamic influences on the arterial wall. New protocols have been developed to identify these risk factors, and a number of candidates have been tentatively identified for evaluation by retrospective study.

Rationale - Geometric Risk Factors

Among the many hypotheses which have been proposed to explain the onset and development of atherosclerosis, a large number invoke hemodynamic factors as being at least partially responsible for the initial formation of these lesions (Gessner 1973; Stehbens 1975; Velican 1978; Nerem and Cornhill 1980). One reason for this inference is the frequent occurrence of atherosclerotic plaques in the vicinity of arterial branches and bends, where unusual flow patterns are expected to develop. Although the role of hemodynamics in the atherogenic process is widely accepted, the most important hemodynamic factor or fac-

Fluid Dynamics as a Localizing Factor for Atherosclerosis
Editors: G. Schettler et al.

tors have not yet been identified with certainty, although almost every hemodynamic phenomenon - abnormally high shear stress, abnormally low shear stress, turbulence or flow disturbance, vortices, flow separation, pressure maxima and minima - has been proposed at one time or another.

All fluid dynamic variables are affected by the geometry of the channel through which the fluid flows. Similarly, the hemodynamic variables which influence the rate and site of the atherosclerotic process are in turn mediated by the geometry of the arterial lumen. For instance, the tendency to separation along the wall of a branch is enhanced as the angle between the daughter and parent vessel increases (Gutstein and Schneck 1967). Thus, if it were shown that flow separation favored atherogenesis, then an abnormally large branch angle could be considered an undesirable geometric feature, placing its possessor at higher risk than a like individual whose branch angle was smaller. This sort of reasoning leads us to the definition of a "geometric risk factor": *an extreme value of a geometric variable, or an uncommon geometric feature, of a susceptible arterial segment which enhances an atherogenic hemodynamic stress.*

Experimental Protocols

Proposing the existence of geometric risk factors is easy; identifying and validating them is harder. The identification of possible geometric risk factors requires an understanding of (i) the response of the arterial wall to its hemodynamic environment and (ii) the dependence of important hemodynamic variables on vascular geometry. These two connected subjects are investigated in our laboratories by selecting at autopsy relatively uninvolved human arterial segments (to date, only aortic bifurcations have been used) exhibiting a realistic geometric variability, passing a life-like pulsatile flow through casts of each vessel's lumen and using a laser Doppler anemometry to measure time-dependent velocities at multiple sites near the walls of each cast. Morphometric and histologic measurements are made at corresponding sites in the original arteries.

The response of the vessel to its hemodynamic environment is examined by seeking correlations between a variety of fluid mechanical quantities derived from the velocity data, and the pathological information on the parent artery. Since the segments selected for study are only minimally diseased, the flow patterns on the lumen have not yet been substantially altered by intruding lesions; thus, such correlations as we might find between morphology and hemodynamics should reflect in some measure the response of a relatively normal vessel to its local fluid mechanical environment. This aspect of our protocol has already been described in some detail (Friedman et al. 1981), but it may be appropriate to summarize briefly the advantages it affords:

1. The geometry of the cast is that of a real, rather than an idealized arterial segment.
2. Mapping is possible between sites at which hemodynamic measurements are made, and sites in the original vessel; this is not the case when idealized experimental or computational models are used.
3. Confounding of results by variability among individuals is avoided, since data obtained from each vessel and its cast are processed individually, rather than being pooled. From the outset, it has been our contention that important information is contained in this variability and that much can be learned from comparisons among individuals; indeed, individual variability is an essential component of the entire geometric risk factor concept.

4. In vitro experimentation permits the use of measurement techniques whose spatial resolution is sufficient for our purposes, but which either cannot be used or whose resolution is too degraded in vivo.
5. Material from humans, the species of ultimate interest, is used.

The dependence of hemodynamic variables on geometry is examined by comparing fluid mechanical measurements in a variety of casts reflecting the normal variability of arterial geometry. The specimens selected for casting and hemodynamic study are chosen to reflect a wide range of geometric variables, and their geometry is preserved in radiographs, photographs, the luminal mold from which the cast is prepared, and the cast itself.

Results

Hemodynamic Influences on the Arterial Wall. As reported earlier for steady flow (Friedman et al. 1980), the anemometer data do not support a role for either separation or turbulence in the localization of lesions at the aortic bifurcation. Transient eddies have been observed on occasion, but they occur only in diastole and are very weak (Mark et al. 1981); turbulence has never been seen. Accordingly, major emphasis has been placed on the relationship between interfacial shear and intimal morphology. Our initial demonstration (Friedman et al. 1981) of a correlation between intimal thickness and shear rates measured in the casts has been supplemented by additional experiments, the results of which will be the subject of a separate report.

Geometric Effects on Mural Shear. The purpose of the analysis which follows is to identify those features of the geometry of the aortic bifurcation which have an important influence on the shear stress at the wall. Consideration will be given here to only the outer wall of the bifurcation, since this is where most of our velocity data have been taken; however, the approach can readily be used to identify the most important geometric determinants of intimal shear at other locations.

The shear at any site along the wall of the branch can be compared to that value which would have been observed at that cross-section were the flow at the site fully developed and parallel. Except for a very short region immediately proximal to the flow divider tip, the cross-sections of the aorta and iliac arteries can be approximated as elliptical; for fully developed pulsatile flow through an elliptical pipe whose axes are a and b, the mean shear rate at the end of axis a is

$$\langle \dot{S} \rangle = \frac{32\langle q \rangle}{\pi a^2 b}\ , \qquad (1)$$

where $\langle q \rangle$ is the mean flow rate through the pipe.

The actual shear rate at a site along the outer wall of the branch will naturally differ from that given by Eq. (1). Almost all of the geometric factors we seek are manifested in that difference; the only exception is local cross-section, whose effect is already included in the equation. To illuminate the effect of individual variations in geometry most clearly, it is convenient to introduce the concept of the "average bifurcation". The longitudinal variation of mean shear rate along the outer wall of the average aortic bifurcation is postulated to be

$$\langle \dot{S} \rangle = \frac{32\langle q \rangle}{\pi a^2 b} P(z/d). \qquad (2)$$

The function P(z/d) measures the influence of the branching geometry of the average bifurcation on the mean shear rate which would be observed at any site were the flow fully developed and parallel at that site. The argument of the function is longitudinal distance, z, normalized by a characteristic diameter, d, of the parent aorta. The effects of the particular geometry of each vessel are seen as deviations between the actual shear rate at a site and that predicted by Eq. (2). The largest deviations are used to identify those geometric features which have the greatest effect on the mean shear rate along the outer wall. Note that this identification process is carried out using data which are realistic with respect to both normal variability and arterial contour.

The paradigm outlined above was applied to data obtained at 52 sites in six casts of minimally diseased aortic bifurcations. Since there is no a priori analytical form for P(z/d), it was estimated by polynomial fitting of $\pi a^2 b\langle\dot{S}\rangle/(32\langle q\rangle)$, termed the "reduced shear rate", vs. z/d (Fig. 1). Because the vessel cross-section and $\langle q\rangle$ change discontinuously on passing from the aorta to the iliac arteries, reduced shear data in the parent and daughters were correlated separately. The

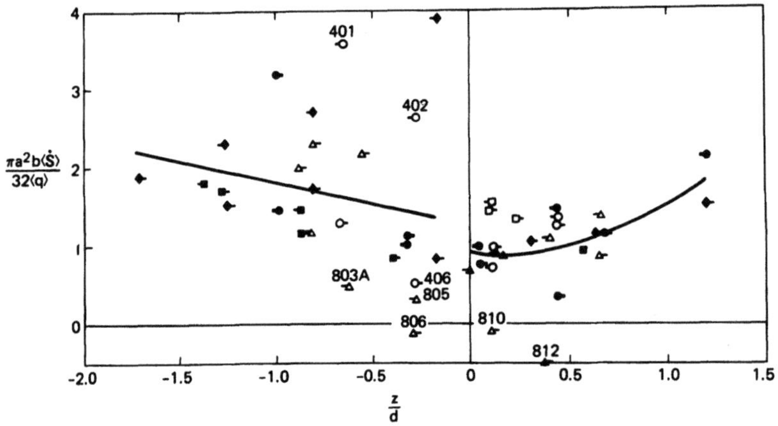

Fig. 1. Reduced wall shear rate vs. normalized longitudinal distance. The values of a and b which define the cross-section at each site are obtained from caliper measurements on the luminal mold; as described earlier (Friedman et al. 1981), the mean rate is obtained from velocity measurements near the wall, and the mean flow rate in the aorta is obtained from centerline velocity measurements in the cylindrical entry to the cast; the mean flows in the iliacs are computed from posttest flow partition measurements. The z-axis is coincident with the approximate centerline of the terminal aorta; z = 0 at the flow divider tip and increases distally. The normalizing diameter is that of a circle whose area is the same as that of the aorta at z = -1 cm; $d = (ab)^{1/2}$. The several symbols, which are tagged to designate the left and right sides of the original vessel, identify data obtained in each of the six casts. The serial numbers of sites identified in Fig. 2 are shown adjacent to the corresponding data point.

The data point at (-0.16, 3.89) lies well above the regression line for two reasons: the site is on an outer wall which is farther from an offset flow divider (v.i.); and it lies so close to the iliac origins that the cross-section is no longer even approximately elliptical. The effect of the latter can be as much as a doubling of the reduced shear rate. The contralateral site at (-0.18, 0.83) lies close to the regression line since, in this case, the effect of flow divider offset opposes that of the non-elliptical cross-section data obtained in the aorta do not support a polynominal of degree greater than unity; a quadratic minimizes the standard error of estimate of the data obtained on the iliac arteries. The fits indicate that the bifurcating geometry causes a broad minimum in reduced shear, the minimum occurring a short distance into the daughter.

Candidate Geometric Risk Factors

For any feature of the vascular geometry to be a possible risk factor, its variability from vessel to vessel must be reflected in significant variations in hemodynamics. Thus, the identification of candidate risk factors can be approached by seeking geometric explanations of the largest deviations from the polynominals in Fig. 1. This was done for the eleven sites at which the reduced shear deviated from the best polynominal fits by more than one standard error of estimate. The wall contours of each branch were obtained from radiographs of the original vessel, photographs of the mold, and coordinate and angle data obtained during the hemodynamic measurements in the cast.

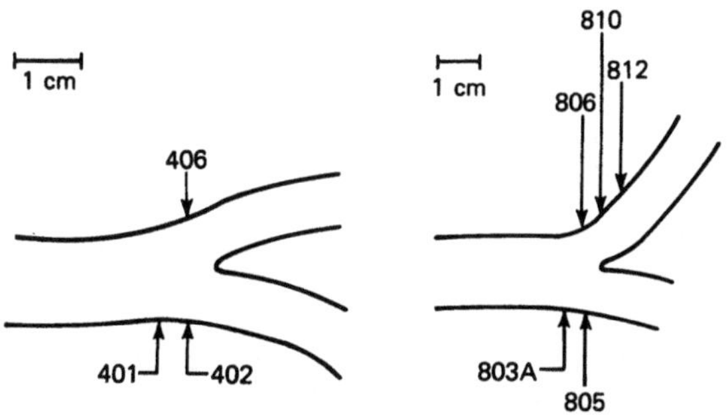

Fig. 2. Tracings of the luminal molds of two vessels used in this study. The shears at the marked sites reflect the geometric features noted in the text: Sites 401 and 402 are in the neighborhood of an incurving aorta; these, and Site 406 precede an offset flow divider tip; Sites 806, 810, and 812 are on the outer wall of a markedly angulating branch; and the shears at Sites 803A and 805 are influenced by the branch angle asymmetry in this vessel

It was found that four specific geometric features were sufficient to explain the deviations at the eleven sites. These features, illustrated in Fig. 2, are described below:

1. A flow divider which is offset from the aortic axis induces considerable asymmetry in the flow field and promotes increased secondary flows. The secondary flows are accompanied by an inordinately high shear stress in the aorta on that outer wall which is *farther* from the tip, and a low shear stress on the opposing outer wall.
2. In most cases, the outer wall of the bifurcation continuously curves away from the aortic axis as z increases. Occasionally, an inward curvature is observed in the aorta, leading to a very gentle intrusion in the flow. Shears in the neighborhood of such intrusions are surprisingly high.
3. When a daughter branch curves away from the parent at an unusually large angle, the outer wall of that branch is exposed to inordinately low shears.
4. In an asymmetric branch, where one iliac artery makes a large angle with the aorta and the outer iliac artery closely follows the aortic axis, the branch geometry approaches that of a parent and side arm. In such cases, sites on the continuing "parent" opposite to the "side arm" can be exposed to relatively low shears. This effect is expected to be sensitive to flow partition and the relative areas of the daughter vessels and the parent.

The fluid mechanical effects of some of these geometric features have been demonstrated earlier, in either numerical studies or experiments in simpler models (Gutstein and Schneck 1967; Lee and Fung 1971; Morgan and Young 1974; Kandarpa and Davids 1976; O'Brien and Ehrlich 1977; Houle and Roach 1981). What is important here is the finding that the variability of these particular features in real arteries is sufficient to cause major variations in the hemodynamic environment at the intimal surface. Of course, the cross-section at a site will also influence the wall shear, through the coefficient of P(z/d) in Eq. (2).

It is possible, on the basis of the somewhat limited data obtained so far, to suggest tentatively a number of candidate geometric risk factors. As the correlation between interfacial shear rate and intimal thickness continues to gain experimental support, it becomes more reasonable that branch angle variations may be geometric risk factors. An offset flow divider tip would appear to be a likely candidate as well, since it is responsible for both inordinately high, and inordinately low mural shears. We are beginning to evaluate these possibilities and others through retrospective studies.

Concluding Remarks

It may not be inappropriate, in closing, to mention two important implications of the existence of geometric risk factors. First, it is reasonable to expect that, as technology moves forward, it will become possible to visualize the human vasculature without pain or hazard to the subject; screening for those geometric factors which have been shown to be closely associated with an accelerated disease process, using noninvasive imaging techniques, is an interesting possibility for the future.

A second, if obvious, implication is that heritable features of arterial geometry may contribute to familiar patterns of atherosclerotic disease. Surely, if we look like our parents on the outside, we are likely to look like them on the inside as well. A number of studies are under way in our laboratories to test this hypothesis.

Acknowledgements. Supported by NIH grant HL21270. The computational support of Stanley Favin is gratefully acknowledged.

References

Anyanwu E, Dittrich H, Jelesijevic V, Druen B, Krefting ER, Hohling H (1981) Coarctation of the aorta. A risk factor in children for the development of arteriosclerosis. Atherosclerosis 39: 367-381

Fishman AP (chairman) (1981) Arteriosclerosis 1981. Report on the Working Group on Arteriosclerosis of the National Heart, Lung and Blood Institute, vol 1. NIH Publication No. 81-2034, p 30

Friedman MH, Bargeron CB, Hutchins GM, Mark FF, Deters OJ (1980) Hemodynamic measurements in human arterial casts and their correlation with histology and luminal area. J Biomech Eng 102: 247-251

Friedman MH, Hutchins GM, Bargeron CB, Deters OJ, Mark FF (1981) Correlation between intimal thickness and fluid shear in human arteries. Atherosclerosis 39: 425-436

Friedman MH, O'Brien V, Ehrlich LW (1975) Calculations of pulsatile flow through a branch. Implications for the hemodynamics of atherogenesis. Circ Res 36: 277-285

Gessner FB (1973) Hemodynamic theories of atherogenesis. Circ Res 33: 259-266

Gutstein WH, Schneck DJ (1967) In vitro boundary layer studies of blood flow in branched tubes. J Atheroscler Res 7: 295-299
Hopkins PN, Williams RR (1981) A survey of 246 suggested coronary risk factors. Atherosclerosis 40: 1-52
Houle S, Roach MR (1981) Flow studies in a rigid model of an aortorenal junction. Atherosclerosis 40: 231-244
Kandarpa K, Davids N (1976) Analysis of the fluid dynamic effects on atherogenesis at branching sites. J Biomech 9: 735-741
Lee J-S, Fung Y-C (1971) Flow in nonuniform small blood vessels. Microvasc Res 3: 272-287
McGill HC, Jr (1978) In: Chandler AB, et al. (eds) The thrombotic process in atherogenesis. Plenum, New York, pp 357-358
Mark FF, Bargeron CB, Deters OJ, Hutchins GM, Friedman MH (1981) Velocity measurements of pulsatile flow through a cast of an asymmetric human aortic bifurcation. In: Van Buskirk WC, Woo SL-Y (eds) 1981 Biomechanics Symposium. Am Soc Mech Engrs, New York, pp 47-50
Morgan BE, Young DF (1974) An integral method for the analysis of flow in arterial stenoses. Bull Math Biol 36: 39
Nerem RM, Cornhill JF (1980) The role of fluid mechanics in atherogenesis. J Biomech Eng 102: 181-189
O'Brien V, Ehrlich LW (1977) Pulsatile flow through stenosed arteries. In: Skalak R, Schultz AB (eds) 1977 Biomechanics Symposium. Amer Soc Mech Engrs, New York, pp 113-116
Stehbens WE (1975) The role of hemodynamics in the pathogenesis of atherosclerosis. Progr Cardiovasc Dis 18: 89-193
Velican C (1978) Hemodynamic stresses and atherosclerosis. Rev Roum Med 16: 3-14

Discussion

C. SCHWARTZ: I have one question which I would like to ask that relates to time frame, i.e. the relationship between flow and the lesions as seen many years after a long period of evolution. Is it possible that the lesions you see at a given point in time are the footprints or the legacy of interactions of many years and that at the time of evolution of the lesions the flow conditions may have differed from those which we now have at this later point in time? In other words, could it be that a much clearer relationship might emerge if earlier lesions or less severely diseased vessels were studied?

FRIEDMAN: It is certainly true that we only get one "look" at the vessel, and we cannot be absolutely certain that the fluid dynamics was similar earlier in time. For this reason, among others, we do use vessels that are only minimally involved. They have small raised plaques at worst, often only Sudan staining.

CARO: I totally agree with you about the importance of geometry; however, the geometry affects the flow and the flow affects the geometry. One should not absolutely lay one's flag to any aspect of this problem - there is a massive interaction of these things. What I wanted to mention very quickly was a paper in Berlin by Dr. Smirnov. He mentioned that he found an association between polygonal cells and lesion occurrence. Are the cells polygonal because they are over lesions or is it something even more complicated? I thus wonder if you and others who are doing this beautiful work would not also look at the endothelial cells? In fact, having said that, it occurs to me that Levesque and Nerem are showing polygonal cells in animals which do not have any lesions.

ANONYMOUS: Of the six studies or casts that you have looked at, four correlated positively with high shear stress, one negatively with shear stress, and a final one was not correlated. For the one that was not correlated and the one that was correlated in a negative way, was there anything special about them? Have you been able to find anything?

FRIEDMAN: The vessel that showed no correlation was run during a period in which we were having difficulty with our equipment, so I do not regard the lack of a correlation in that case as particularly significant, although I cannot prove that was why we did not get one. The feature that seemed most striking about the one that correlated negatively was that the branch was very symmetric and exhibited a relatively large branch angle more evenly divided between the two daughter vessels than any of the others. In selecting the next couple of aortic bifurcations to look at, we have chosen vessels much like that one, to see whether or not we again find negative correlations.

BLACKSHEAR: Did you attempt to correlate lesions with the thickness of the unvascularized media?

FRIEDMAN: We did not do that, specifically. We have looked at full medial thickness. In the first three vessels that we did, we got good positive correlations between medial thickness and shear rate.

SCHMID-SCHÖNBEIN: I am not sure that I understand what you actually mean by pulse shear rate. How does this relate, for example, to pressure and/or flow pulsatility?

FRIEDMAN: Unfortunately, in our experiments we do not use elastic casts. We use rigid vessels through which we force a pre-determined flow wave. It is done with a cam-driven piston pump, and it gives us a normal in vivo flow. Physiological pressure is not modeled in our experiment. The pulse rate is defined just as I think you would define pulse pressure. It is the algebraic difference between the maximum shear rate and the minimum shear rate at the site; the latter shear rate is almost always negative. If the shear profiles were similar from site to site, then the ones that had the highest pulse shear rates would have the highest dS/dt, where S is shear rate. However, I have not looked to see what the maximum rate of increase of the shear rate is.

The Drainage of Transmural Flow and the Consequences of Its Insufficiency

H. Jellinek[1]

It has been shown that permeability disorders occurring in the early phase of vascular injury play a decisive role in the development of mural lesions. Deposition of plasma in the vascular wall was demonstrated by various tracer substances in experimental hypertension (Jellinek et al. 1969; Olsen 1969; Giacomelli et al. 1970; Hüttner et al. 1970) and hypercholesterolaemia (Adams et al. 1968; Gore et al. 1972; Veress et al. 1972) but it has been supposed to play a part in the development of hypoxic vascular lesions as well (Astrup et al. 1967; Kjeldsen et al. 1968). Systemic hypoxia may, in itself, lead to the development of arteriosclerotic lesions (Garbarsch et al. 1969; Helin et al. 1969). And it enhances atheromatosis in cholesterol-fed animals (Astrup et al. 1967; Kjeldsen et al. 1968). A temporary hypoxia causes irreversible changes associated with a severe permeability disorder in muscular-type small arteries (Kerényi and Jellinek 1972).

Experimental models were developed in animals to study the problems of transmural transport in vessels. The basic experimental system was the *hypoxia model*. In this model the exalted permeability was demonstrable within the first 48 h by the appearance in the vessel wall of the intravenously administered *colloidal iron*. The latter was visualized in histological preparations by the Prussian-blue reaction. In the first hours colloidal iron was seen only in the intima, while at 48 h it appeared in the full width of the vessel wall. During the following ten days the amount of colloidal iron in the vessel wall decreased and finally the transmural transport became normal again.

Similar phenomena were observed after a damage causing *local necrosis* in the vessel wall. Following the application of concentrated acid to the aorta wall an extensive exaltation of permeability developed for a period of 48 h, as revealed by the appearance of colloidal iron in the full width of the vessel wall.

In both systems the presence of colloidal iron was demonstrable in the vessel tissues not only by light-, but also by electron microscopy. Iron granules were easily seen in uncontrasted preparations. They were present partly within the endothelial cells, partly in the enlarged junctions. In both types of models the damaged endothelial cells contained, prior to their lysis, large numbers of iron granules. Iron was, however, present also in the subendothelial space, in the smooth muscle cells and in the intercellular space in the media as well as in the adventitial fibroblasts.

---

1 From July 1, 1981 - June 30, 1982. Visiting Professor at the Department of Pathology and Herzinfarktforschungs-Institut of the University of Heidelberg

Fluid Dynamics as a Localizing Factor for Atherosclerosis
Editors: G. Schettler et al.

The same observations were made also in an arteficial *hypertension model*. The extent of the appearance of colloidal iron was proportional to the increase of the tension as shown both by light and electron microscopy.

Exaltation of permeability was demonstrated also by *another tracer* i.e. by Lipofundin "S" (Braun, Melsungen), a soybean extract plasma supplement. In hypoxia experiments Lipofundin was demonstrated light microscopically in preparations stained for fat. If administered simultaneously and in equal amounts with colloidal iron, both tracers were seen to be present in the endothelial cells in the early phases of damage. Both Lipofundin and iron were demonstrable not only by light-, but also by electron microscopy.

Sections by light microscopy showed that the endothelial cells located along the luminal side of the aortic wall were filled with a sudanophilic substance. This substance was either distributed focally, or occurred in several neighbouring endothelial cells. The nucleus, in lateral position, could be easily distinguished from the sudanophilic Lipofundin substance filling the endothelial cells.

In *semi-thin sections* stained with toluidine blue, already is aid lymph vessel are seen to associate directly with the aortic wall. Capillaries containing red blood cells are easily visible near the lymph vessels. In other semi-thin sections, also stained with toluidine blue, globules taking on a greenish stain are apparent inside the distended lymph vessels. Apart from lymph vessels closely associated with the external muscular layer of the aorta, several blood vessels filled with erythrocytes are visible.

The *electron micrographs* indicated that Lipofundin occurs both in the recesses and in the cytoplasm of endothelial cells (Fig. 1). The iron granules were attached either to greater details of Lipofundin, or only to its surface, imparting to it a dark contour. Iron granules were seen inside endothelial cells also independently of Lipofundin. In places the iron tracer surrounded the Lipofundin droplets in a cap-like manner.

Lipofundin, surrounded by iron granules and with cap-like deposition of iron at one side, was also seen in the muscle cells of the media (Fig. 2). A similar iron-marked droplet of Lipofundin was encountered inside the wall of a muscular-type small artery that branched from the ligated aortic segment. In the adventitial area, droplets of Lipofundin localized between the collagen fibres (Fig. 3) and in some adventitial fibrocytes. Many of these droplets were surrounded by granules of the iron tracer.

The passage of Lipofundin can be still more easily followed in aortic lymph vessels. A section including the detail of a periaortic lymph vessel - the area near the external collagen fibres - shows entrance of a Lipofundin droplet into a cell via pinocytotic activity. The presence of Lipofundin inside the cell, and its passage across the cytoplasm are easily visible; part of the Lipofundin substance is surrounded by iron (Fig. 4). On reaching the luminal border, Lipofundin is released from the cell into the lumen (Fig. 5). The static picture permits functional evaluations because the Lipofundin and the iron can penetrate only from the doubly ligated aortic "sac" into the adventitia and the adventitial lymph vessels. In some places only the iron granules are seen inside the pinocytotic vesicles of the lymph vessel endothelium, which are more numerous at the luminal than at the other side.

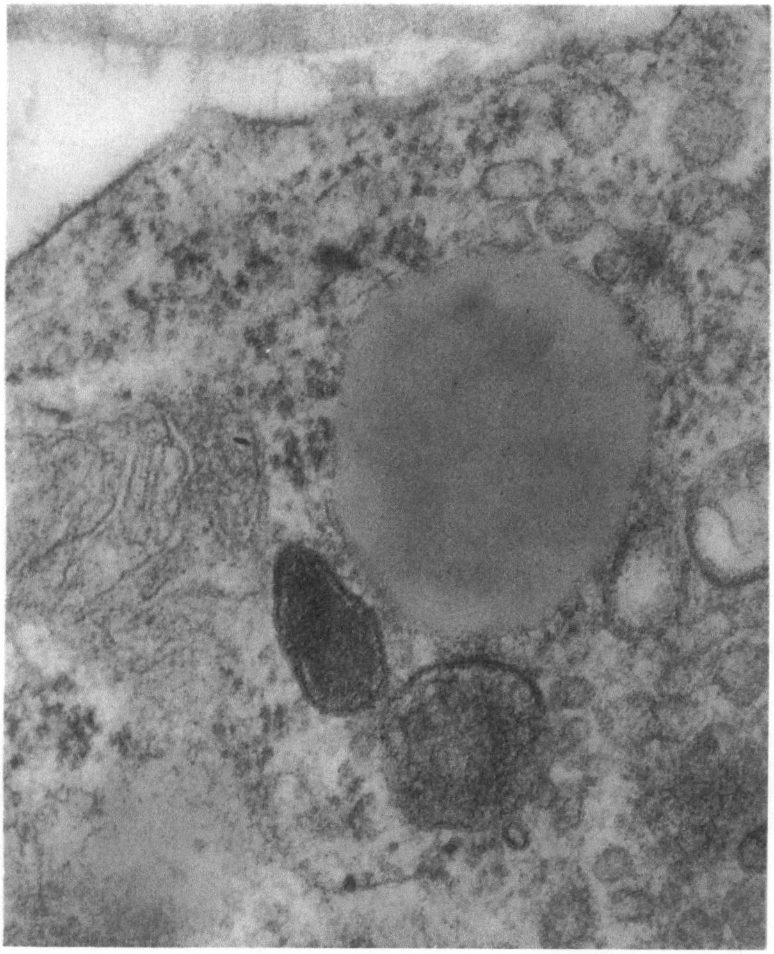

Fig. 1

The localization of Lipofundin between collagen fibres, indicated that Lipofundin droplets (originally 1, μ in diameter) occasionally merge to form larger pools. In the adventitial lymph vessels we observe the degradation of Lipofundin into small fragments. These are later emptied into the lumen.

In the adventitial drainage the lymph vessels seem to play an important role. This has been demonstrated in the hypoxia and high-lipid diet experiments. In hypoxia macrophages with phagocytosed smooth muscle cell debris were seen in the lumen of lymph vessels while high-lipid diet caused dilatation of these vessels and appearance of lipid-like matter in the lumina.

Thus these experiments have revealed the pathway of material transport and pointed to the importance of the lymph vessels.

The plasma substances depositing in the subendothelial space in consequence of the permeability increase undergo a characteristic crystallization and finally a hyaline transformation, the lipids and lipid-

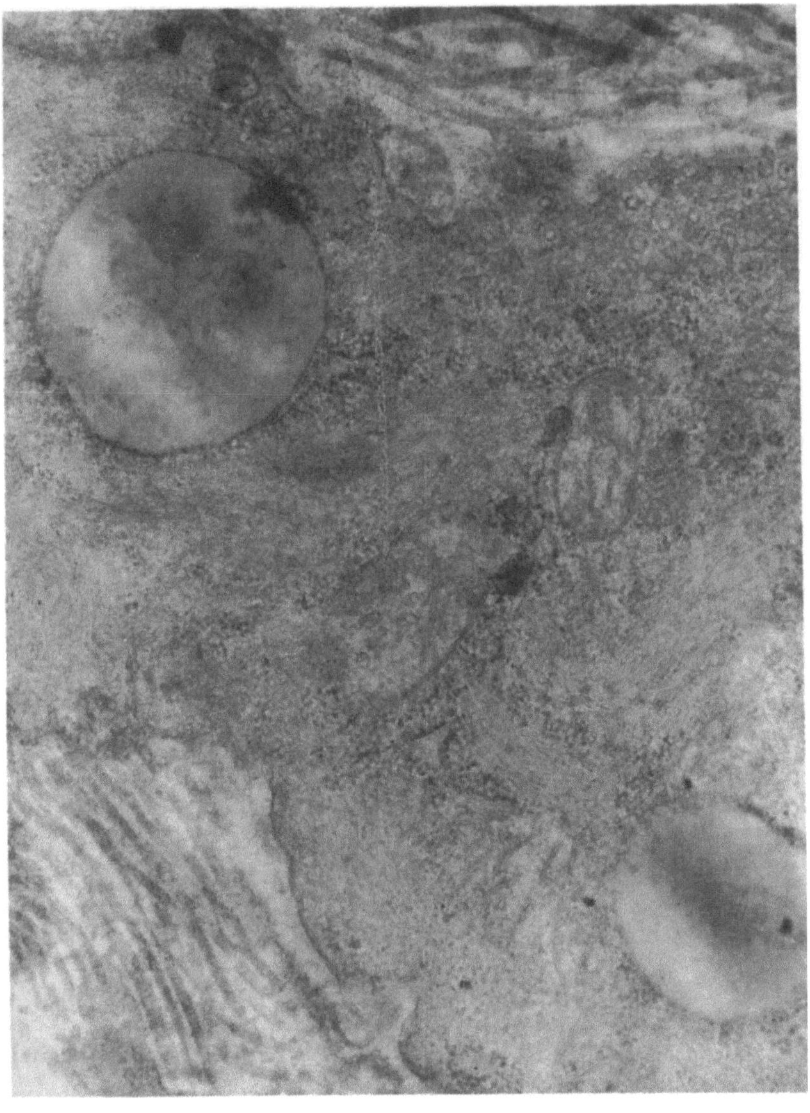

Fig. 2

like substances can first be detected ultrastructurally in the hyaline tissue thus formed.

In other cases the plasma deposit ion becomes phagocytized resulting in intimal proliferation. The latter progresses to a cellular sclerotic plaque, in which lipid-like substances will later appear as a result of degenerative phenomena.

The administration of colloidal iron makes possible to follow the iron granule transport from the intima to the adventitia under various vascular damages. During this process the iron was shown to enter the smooth muscle cells, where it was easily demonstrable. This suggested that also plasma entered these cells and accordingly intracellular fibrin was detected in electron microscopic preparations. Under these

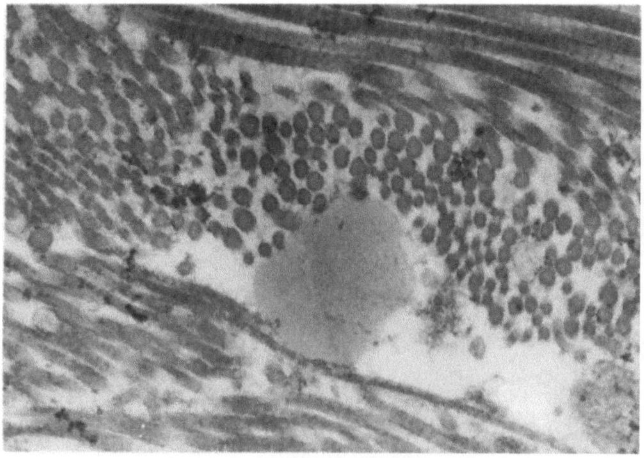

Fig. 3

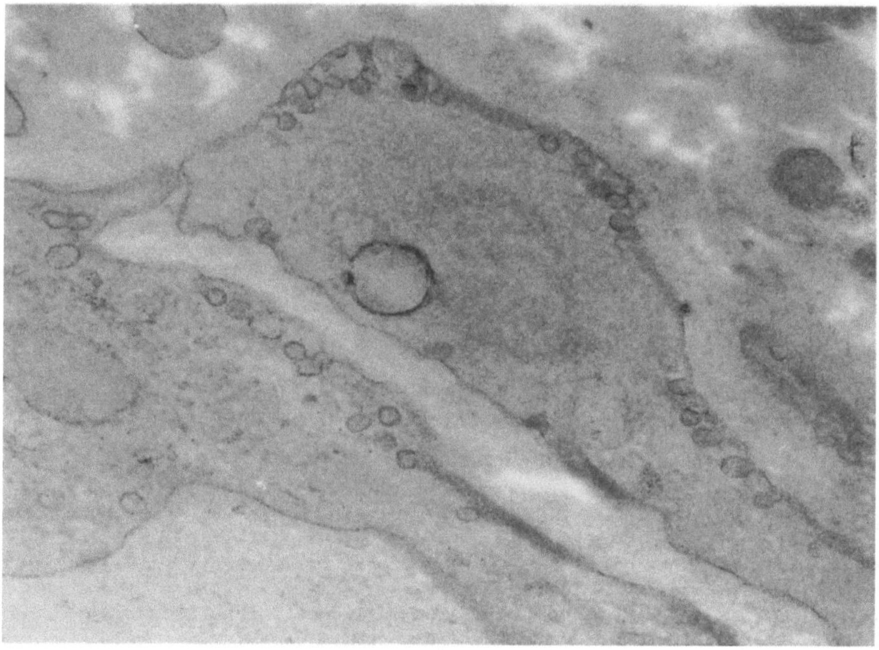

Fig. 4

conditions the smooth muscle cells may necrotize. This process may initially manifest only in some areas of the cell and produce the picture of edematic or osmiophilic necrosis.

The deterioration of single muscle cells acts towards the increase of the regeneration capacity of the preserved cells, which again results in cellular plaque formation, followed by secondary degeneration and lipid deposition.

Thus actually all factors affecting vascular permeability appear to result finally in sclerotic lesions: hyalinosis, phagocytosis of plasma

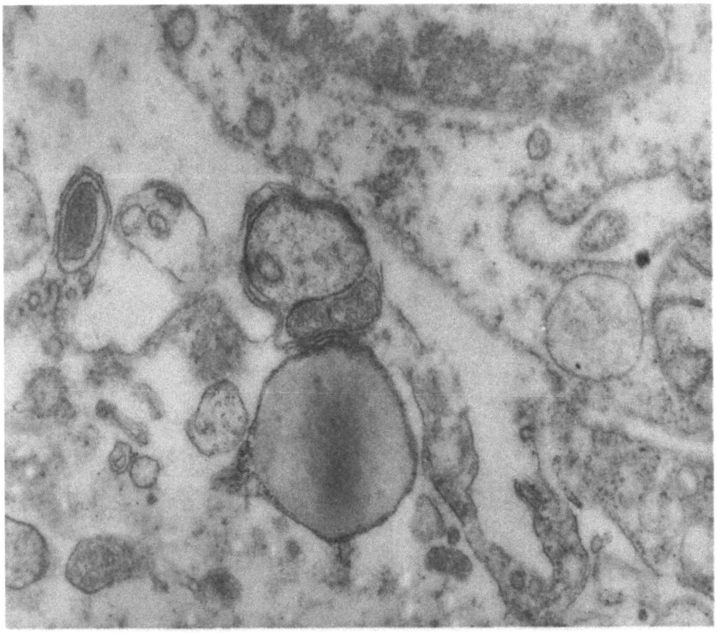

Fig. 5

proteins and cells proliferation consequent upon muscle cell degeneration equally result in vascular plaque formation and cellular degeneration.

Apart from the four kinds of vascular injury reproduced in experimental models, many other factors may be held responsible for vascular permeability changes, because the blood circulation participates in fact in pathological processes affecting the body. The complex effect of various factors can account for the rapid progression of arteriosclerosis.

References

Adams CWM, Virâg S, Morgan RS, Orton C (1968) J Atheroscler Res 8: 679

Astrup P, Kjeldsen K, Wanstrup J (1967) J Atheroscler Res 7: 343

Garbarsch C, Matthiessen ME, Hellin P, Lorenzen I (1969) J Atheroscler Res 9: 283

Giacomelli F, Wiener J, Spiro D (1970) Am J Path 59: 133

Gore J, Miyaski T, Kurozumi T, Hirst A (1972) Am J Path 66: 47

Helin P, Lorenzen J, Garbarsch G, Matthierse M (1969) J Atheroscler Res 9: 295

Hüttner I, Mare R, Jellinek H (1970) Lab Invest 22: 501

Jellinek H, Nagy J, Hüttner I, Bâlint A, Kroczê A, Kerênyi T (1969) Br J Exp Path 50: 13

Jellinek H (1975) Arterial lesions and arterioscler. Plenum Press, London New York

Jellinek H, Kerênyi T, Detre Z (1979) Exp. art. injuries of the rat and arterioscler. Atlante Grafico Della Arterioscler. Vol. 2. Skema, Bologna

Kerênyi T, Jellinek H (1972) Exp Molec Path 17: 1

Kjeldsen K, Wanstrup J, Astrup P (1968) J Atheroscler Res 9: 835

Olsen F (1969) Acta path microbiol scand 75: 527

Veress B, Bâlint A, Jellinek H (1972) Morph Igasz Orv Szemle 12: 178

Discussion

CARO: I think that you make an extremely interesting point that the change of permeability or change of the outflow has profound effects on the arterial wall.

ANONYMOUS: What is the difference between Dr. Schwart's observation that ferritin-marked albumin will stick to the arterial wall, whereas in your system where you use the colloidal iron you do not observe this binding? Is this purely a mechanism which can be explained by electrostatical charges or is there any other difference?

JELLINEK: First, in the experiment designed by Dr. Schwartz, cationized ferritin was used which may differently react with the endothelium than the colloidal iron. We do not have any information about the electrical charge of the colloidal iron complex. Besides, ferritin has a molecular weight of 400,000 - 500,000 while the colloidal iron is 40,000 - 80,000 molecular weight, which could suggest different transport mechanisms.

ANONYMOUS: Would this mean that the colloidal iron will have the possibility to migrate through the wall?

JELLINEK: Normally a very low permeability for colloidal iron can be detected in the aorta. If the endothelium is damaged, the entrance of the colloidal iron increases dramatically. In electrophoretic studies, the colloidal iron always moved together with albumin and α_1 globulin. Therefore we considered the entrance of the colloidal iron as a marker of "plasma insudation". The entrance of this tracer can not be regarded as migration. It is actively taken up by endothelial cells by micropinocytosis. Then a part of it enters the endothelial cytoplasm, while a large amount of the tracer is delivered into the subendothelial space. Then it moves via stomata of elastic lamellae toward the media where again part of it is taken up by medial smooth muscle cells, etc.

SEED: I would like to ask you whether the severity of the hypoxia that you used in these experiments was comparable to the severity of hypoxia that one finds during life?

JELLINEK: I have not had time to provide the details. Actually we performed two types of experiments. In the first one a local transient anoxia was induced which is very severe and seldom occurring in life. In the other experiment the animals were kept in an hypobaric atmosphere for 4 weeks, and this experiment can bring much more relevance to situations occurring in human life.

SEED: In the chronic experiments, what was the partial pressure of oxygen in the blood? Chronically maintained hypoxia is a situation which occurs clinically in an enormous number of people with lung disease. I am not aware of any good pathological evidence that these people have an accentuated vulnerability to arterial disease. I wonder if you have any comment about that.

JELLINEK: That is a very good question and I must tell you that we have tried to answer it several times in my department. Hypoxia is a pathogenetic factor known to aggravate arterial lesions. However, in human hypoxemia due to lung disease or chronic heart failure, it is very difficult to assess to what extent hypoxemia is responsible for the arterial lesions and to what extent to the other known risk factors.

SPRAGUE: I would like to provide a clarification on the date regarding ferritin. There is a very big difference between the colloidal iron used in these experiments and the ferritin used in the study by Dr. Schwartz and me. This is the fact that we are dealing with positively charged ferritin, a cationized ferritin which we are using to preferentially adhere to the surface of the endothelium for very special reasons. This is to study whether it causes a redistribution of surface charges. We are interested in endocytosis and in whether the uptake of these positively charged molecules might affect the endocytosis of molecules. So we are looking at something very different than the effects of altered permeability.

Fluid Mechanical Factors in Macromolecular Transport by the Arterial Wall[1]

S. Chien[2]

There is considerable evidence that hemodynamic factors may alter the arterial wall transport process and contribute to atherogenesis [1,2, 3]. The purpose of this presentation is to summarize some of our investigations on macromolecular uptake by the arterial wall as affected by oscillatory mechanical disturbance and pressure variations and on the regional variation of uptake in relation to ultrastructure and hemodynamics.

1. Effects of Oscillatory Mechanical Disturbance on Macromolecular Transport by Arterial Wall

Transport of ^{125}I-albumin by isolated segments of canine common carotid arteries was studied in vitro at zero transmural pressure [4]. Sinusoidal oscillatory variations in length (peak change 4%) for 15 minutes at frequencies of 5 and 10 Hz caused 40% increase in ^{125}I-albumin uptake, and also a 30% increase in the apparent luminal surface area. Changes in duration and frequency of oscillation indicate that the total number oscillation (= frequency x duration) was the critical parameter in causing these effects. The increase in apparent luminal surface area was correlated with regional flattening of the internal elastic lamina and the overlying endothelial cells, as demonstrated by transmission and scanning electron microscopy. Endothelial vesicles were counted with the aid of ruthenium red as a post-fixation extracellular marker. The ratio of unstained free vesicles to total vesicles averaged 0.083 in the control state and decreased slightly to 0.070 after oscillation. Although the decrease in free vesicle population indicated an acceleration of vesicle diffusion, computations based on our theoretical model [5,6] showed that the resulting increase in vesicle flux was negligible. The increase in ^{125}I-albumin uptake by the artery following mechanical oscillation is mainly attributable to the increase in apparent luminal surface area.

2. Effects of Pressure Variations on Macromolecular Transport Across Vessel Wall

Effects of changes in arterial pressure (P) on macromolecular transport were studied in canine common carotid arteries in situ. An increase in P led to an enhancement of ^{125}I-albumin permeability, which was 1.04 ± 0.12, 1.29 ± 0.19 and 2.20 ± 0.73 (in 10^{-7} cm/sec, mean $\pm$ S.E.M.) for P = 0, 100 and 200 mmHg, respectively. The ratio of free to total vesicles (determined in the peripheral zone of endothelial cells by using ruthenium red as a postfixation extracellular marker)

1 Supported by U.S.P.H.S. Grant HL19454 from NHLBI and NSF Grant ENG7822101
2 Presented by Professor S. Weinbaum, City College of New York

Fluid Dynamics as a Localizing Factor for Atherosclerosis
Editors: G. Schettler et al.

was 0.090 at 0 mmHg, and it decreased to 0.074 at 100 mmHg and 0.057 at 200 mmHg. According to our theoretical model for transendothelial vesicle diffusion [5,6], these values indicate that a rise in P from 0 to 200 mmHg led to an increase in vesicle flux by only 4%. Increase in P also caused flattening of endothelial cells and a reduction of transendothelial diffusion distance; theoretical computation shows that this can lead to an increase in permeability by 20-30%. The role of geometric hindrance at the vesicle neck in macromolecular transport was studied by using Ag and Au colloids (diameters = 9, 13, 15, 19, 25 and 35 nm; coated with polyethylene glycol). At 100 mmHg, only 9 and 13 nm particles were found within endothelial vesicles and in subendothelial space; at 200 mmHg, particles as large as 25 nm became detectable. Therefore, at physiological P of 100 mmHg, macromolecules 15 nm or larger, e.g. the 19-22 nm low density lipoprotein (LDL), would be excluded from the receptor-independent vesicle transport. In hypertension, vesicular loading of larger size particles (19-25 nm), in addition to the enhanced permeability due to reduction of transendothelial diffusion distance, may cause an enhanced entry of LDL into arterial wall and contribute to atherogenesis.

3. Regional Variations in Macromolecular Uptake and Ultrastructure in Aorta [7]

Following the localized removal of endothelium from the rabbit aorta with the use of a specially designed catheter, Evans blue labeled albumin was injected intravenously and the animals were sacrificed 1 hr after the injection. The aorta was subjected to perfusion fixation with 3% glutaraldehyde at 100 mmHg. The normalized average tissue dye concentration (U), determined by using a fluorescence microscope and a video-digitizer, was higher in the dorsal than the ventral aspect of the aortic arch. The U values in both aortic arch specimens were higher than those in the other regions, with a trend of progressive decrease toward the more distal segments of the aortic tree. Electron microscopic examinations show that the dorsal aortic arch has a disorganized endothelial surface, nearly round endothelial cell nuclei which are not oriented, and a large subendothelial space containing many smooth muscle cells. In contrast, the ventral aortic arch and the more distal regions have a more regular endothelial surface, relatively more oval nuclei which are oriented with blood flow, and a narrower subendothelial space containing fewer smooth muscle cells. Therefore, the regional difference of dye uptake can be correlated with that of morphological characteristics. These results, together with previous reports on the variations of hemodynamic conditions along the aortic tree [8,9,10], support the theory that regional blood flow characteristics may play a significant role in affecting local ultrastructure and macromolecular uptake in the arterial wall.

References

1. Fry DL (1973) Responses of the arterial wall to certain physical factors. In: Atherogenesis: Initiating factors. Ciba Foundation Symposium 12 (new series). Associated Scientific Publishers, Amsterdam, pp 93-125
2. Nerem RM, Guyton JR (1980) Hemodynamics and the arterial wall. Proceedings from a Specialists Meeting sponsored by National Science Foundation, University of Houston and Baylor College of Medicine. University of Houston, Tx
3. Caro CG (1981) Arterial fluid mechanics and atherogenesis. Recent Adv Cardiovasc Dis (Nat. Cardiovasc. Center, Suita, Osaka, Japan), vol II [Suppl], pp 6-11

4. Chien S, Lee MML, Laufer LS, Handley DA, Weinbaum S, Caro CG, Usami S (1981) Effects of oscillatory mechanical disturbance on macromolecular uptake by arterial wall. Ateriosclerosis 1: 326-336
5. Weinbaum S, Chien S (1980) Vesicular transport of macromolecules across vascular endothelium. In: Gross JF, Popel AS (eds) The mathematics of microcirculation phenomena. Raven Press, New York, pp 109-131
6. Chien S, Weinbaum S (1981) Vesicle transport in arterial endothelium and the influence of mechanical factors on macromolecular permeability. J Biomech Eng 103: 186-196
7. Thurn AL (1982) Effects of endothelial injury on macromolecular transport in the arterial wall. Ph.D. Dissertation, Columbia University
8. Schultz DL (1972) Pressure and flow in large arteries. In: Bergel CH (ed) Cardiovascular fluid mechanics, vol 1. Academic Press, New York
9. Caro CG, Pedley TJ, Schroter RC, Seed WA (1978) The mechanics of the circulation. Oxford University Press, Oxford
10. Stein PD, Sabbah HN, Anbe DT, Walburn FJ (1979) Blood velocity in the abdominal aorta and common iliac artery of man. Biorheology 16: 249-255

Discussion

ANONYMOUS: For the same ratio of attached to non-attached vesicles, could it not be that there is a pharmacology of these such that there is a faster and slower movement which would not change the ratio?

WEINBAUM: I believe that the vesicle population, that which functions passively and which does not require metabolic energy, has to be driven by Brownian motion. It is distinctly different from the vesicle population that is either involved in engulfing particles or certainly the ones which are receptor mediators because those vesicles are much, much larger.

ANONYMOUS: In the calculation of the proportion of the vesicles not attached, are you certain that they are not attached to the opposite side, since the ruthenium red technique would only tell you if they were attached to the luminal side?

WEINBAUM: We know that, where the junctions are leaky, the ruthenium red will go right down into the underlying tissue and spread beneath the endothelium so that we can get a counting of the attached vesicles on the tissue's surface.

ANONYMOUS: How can you decide that there is a change in transjunctional permeability as opposed to an increase in vesicular transport?

WEINBAUM: The one thing I can say with certainty is the maximum rate of increase in permeability due to the vesicle population. This is because I know what the attachment time is and I know what the diffusion time is, and if I were to reduce the diffusion time to zero and the process becomes rate-limited by attachment time, I can then tell you what the permeability is. What the theory tells me is that, when the ratio is 11.3 to 1, the maximum increase in transport would only be on the order of 8%. I could not increase it beyond that unless I was introducing some chemical process which changed the vesicular attachment time. Thus it is impossible to get a two hundred percent increase in uptake using vesicle transport through a passive mechanism like we have. Now I do not know whether endocytosis is involved in trans-endothelial permeability or not. Locally, I know one can get a two-fold increase when cells are growing, but that will not give the numbers which people have measured for increased permeability.

SPRAGUE: Accepting those calculations that you have given us as far as attachment in available vesicles, I think there is another process to consider. This is the fact that you could alter the number of vesicles that are actually involved in the transport of albumin without necessarily altering the rate. I think if you look at some work that was done quite a long time ago by Dr. Schwartz in terms of the uptake of ferritin particles in Evans Blue areas versus white areas, what he found was that a relatively small number of vesicles actually participated in the uptake compared to the number that were available. Thus you have to look at changes in the biochemical surface of the vesicles that might enhance the number that are involved in the uptake of albumin and also the quantity that each vesicle might transport.

WEINBAUM: The question you raise is one about which we have thought a lot, but we do not interpret in that way. If you are talking about a large marker molecule like ferritin, then you have a problem with the filling of the vesicle, and many of the vesicles may be involved in transport and not available because you never fill them. However, if you are talking about a much smaller molecule, like albumin, the problem of filling, I believe, is not a significant one. There may be barriers at the mouth of the vesicles and there is the glycocalyx; however, I am not aware from our own studies of the existence of diaphrams or diffusion barriers in the arterial endothelium. Of course, this has to be documented. There is a glycocalyx there; but from what we see, it is very loose.

The Role of Hydrodynamic Factors in Platelet Activation and Thrombotic Events: The Effects of Shear Stress of Short Duration

L. J. Wurzinger, R. Opitz, P. Blasberg, H. Eschweiler, and H. Schmid-Schönbein

Introduction

More than a century after Rokitanski's [14] incrustation hypothesis of atherogenesis, good evidence has accumulated that white, platelet rich thrombi are not only involved in the complication of atherosclerosis [6] but also in its genesis. Ross's detection of a smooth muscle cell proliferation stimulating factor, released from platelets stresses the paramount role of platelets in the development of the disease [15].

The initiation of white thrombi requires quite different fluid dynamic boundary conditions than red stasis thrombi. White thrombi are only formed under conditions of maintained rapid blood flow - an observation almost as ancient as Rokitanski's incrustation hypothesis [21]. The preferential localization of atheromatous plaques at branchings and bends in the arterial tree, i.e. zones of disturbed and regionally enhanced blood flow, suggests influence of hydrodynamic factors in atherogenesis too.

In the present paper we concentrate on biological effects of hydrodynamic forces on platelets and the coagulation system. There is ample evidence that hydrodynamic forces damage and/or activate platelets, i.e. produce sharp change, and release, and enhance platelet deposition upon surfaces. Some of these studies, however, employ tube flow and other hard to define flow conditions [4,7,12,17,20]. Since the effective concentration of platelets in tube flow (Blasberg et al., this volume) is a function of the distance from the wall and the presence of red cells, only a rough estimate of the number of platelets and the time they are exposed to a calculated shear force can be undertaken.

Important studies on platelet activation by shear forces have been done using Couette viscometers, shearing blood cell suspensions in a gap between two concentric cylinders [1,2,16]. In this flow field a homogeneous distribution of a precisely adjustable shear force throughout the entire sample volume is ensured and there are no significant phase separation effects.

Unfortunately the earlier Couette experiments expose platelets to high shear stress for several minutes. In vivo, however, shear stresses beyond 5 N/m^2 can only be expected regionally during systolic peak flow for milliseconds (ms) (Giddens et al. and Liepsch et al., this volume).

While important in our understanding of platelet microrheology, the earlier shear experiments can only be accepted as relevant for the actual platelet behaviour if shear stresses acting for milliseconds can elicit similar effects. We therefore carried out experiments in a novel Couette viscometer, which enabled us to expose platelets to shear fields up to 21.400/s for 7 to 700 ms [9]. In aggregometer stud-

Fluid Dynamics as a Localizing Factor for Atherosclerosis
Editors: G. Schettler et al.

ies we have learnt that human platelets are extremely sensitive to changes in temperature [13], and that citrate anticoagulation depresses platelet reactivity to biochemical stimuli [19]. As a consequence of this, we used heparin anticoagulation and a protocol, ensuring strict isothermia (37°C).

Methods

Platelet rich plasma (PRP), prepared from whole blood, anticoagulated with 5 U/ml heparin (Liquemin[R]) was used in our experiments. All steps of the experimental procedure were carried out at 37°C. Platelet count was adjusted to $2.5 \times 10^5/mm^3$ by dilution with autologous PRP.

PRP was transferred to a plastic bag in a thermostatized pressure container, connected to a pump. From the bag silicone rubber tubes lead into the shear apparatus (for details see [9]). The latter consists of two concentric cylinders, with the inner cylinder rotating, thus generating shear rates of up to 21.400/s or 255 N/m^2 at a viscosity of 1.2 mPa.s (Fig. 1).

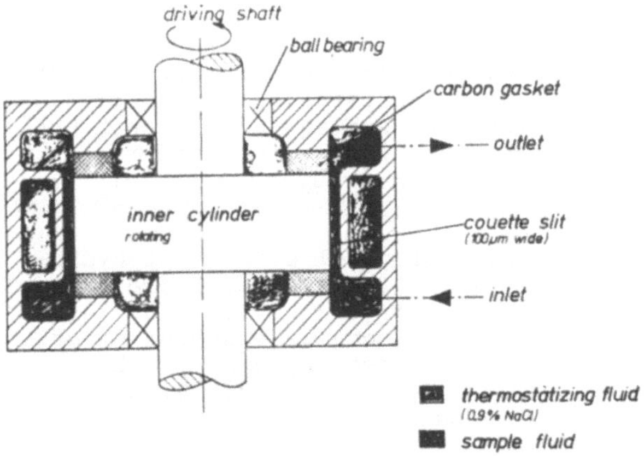

Fig. 1. Schematic drawing of the flow through Couette viscometer of Heuser and Opitz [9]

The test fluid, in our case heparinized PRP, is pumped through the Couette slit in axial direction, which sends particles on a spiral pathway through the slit. The flow rate of the PRP, generated by the pump operating on the pressure container with the PRP bag, determines the time a platelet needs to traverse the high shear region in the Couette slit. The shear force in turn is determined by the rotational speed of the inner cylinder.

During a set of experiments done on one PRP batch for a given exposure time, 4 different shear stresses were applied including a fifth sample, pumped through the slit with stationary inner cylinder, serving as unsheared control. The shear experiments were started 50 minutes after blood withdrawal and finished within 15 minutes. Air contact of PRP was avoided by priming the whole apparatus with saline. After each single experiment (1 shear stress, 1 exposure time) the apparatus was rinsed with distilled water first and then with saline.

The sheared samples were processed for further investigation immediately after passing the apparatus. The following platelet parameters were assayed:

1. β-Thromboglobulin (β-TG) as indicator for release of platelet granular constituents. PRP was processed for β-TG RIA with a commercially available kit (Amersham) according to Ludlam [10]. The results obtained with sheared PRP were expressed as percentage of the concentration measured in an aliquot of three times deep frozen PRP, after subtraction of the concentration measured in the unsheared control.
2. Lactic dehydrogenase (LDH) as cytoplasmic marker enzyme indicating platelet destruction. LDH activity in the supernatant of samples and in frozen-thawed preparations was assayed within 2 hours after shear exposure, using a kit from Boehringer/Mannheim. Percent LDH liberation was calculated accordingly to β-TG release.
3. Stypven time was taken for platelet procoagulant activity, in other terms platelet factor 3 (PF 3) activity. This test is based on activation of factor X by Russell's viper venom (RVV). In the presence of factor Va and ionized calcium, conversion of prothrombin to thrombin by factor Xa, and thus initiation of coagulation critically depends on phospholipid surface provided by platelets [8]. Shortening of the Stypven time is thus taken as a sign of the exposure of procoagulant platelet phospholipids.
4. Platelets were counted phase contrast microscopically.
5. Platelet aggregation (PA) was monitored in a rheoaggregometer [5] 30 minutes after shearing, by adding ADP in 5 μM final concentration.
6. Transmission electron microscopy (TEM) was performed after glutaraldehyde fixation, OsO_4 postfixation, acetone dehydration and araldite embedding according to Morgenstern [11]. After staining with lead citrate and uranyl acetate, ultrathin sections were viewed in a Zeiss EM 9 microscope. For TEM ACD-anticoagulated PRP of a pH of 7.4 was used. The other parameters were checked and found to be within the range of those, obtained with heparinized PRP.

Results

Using the afore described techniques to expose heparinized PRP for 7, 27, 113, and 700 ms to 4 different levels of shear stress (57, 108, 170, 255 N/m^2), we obtained the following results (Figs. 2 and 3):

1. The parameters LDH liberation, β-TG release, and PF 3 availability, are obviously influenced by the duration as well as the magnitude of shear forces (Fig. 2).
2. The absolute magnitude of shear, however, is more effective than its duration with regard to these parameters, indicative of platelet stimulation and/or damage (note log-scale of exposure time axis in Fig. 2!).
3. PF3 availability appears to be the most sensitive indicator of shear induced alteration of platelets.

Table 1. % Reduction of Stypven time after shear exposure

τ \ t_{exp}	7 ms	27 ms	113 ms	700 ms
57 N/m^2	0.1 ± 11	10 ± 14	27 ± 9**	52 ± 21**
108 N/m^2	3 ± 12	26 ± 7*	33 ± 11**	68 ± 17**
170 N/m^2	13 ± 15*	36 ± 11**	55 ± 5**	84 ± 8**
255 N/m^2	50 ± 21**	63 ± 5**	75 ± 4**	

$x \pm$ S.D.; * $p < 0.05$; ** $p < 0.01$

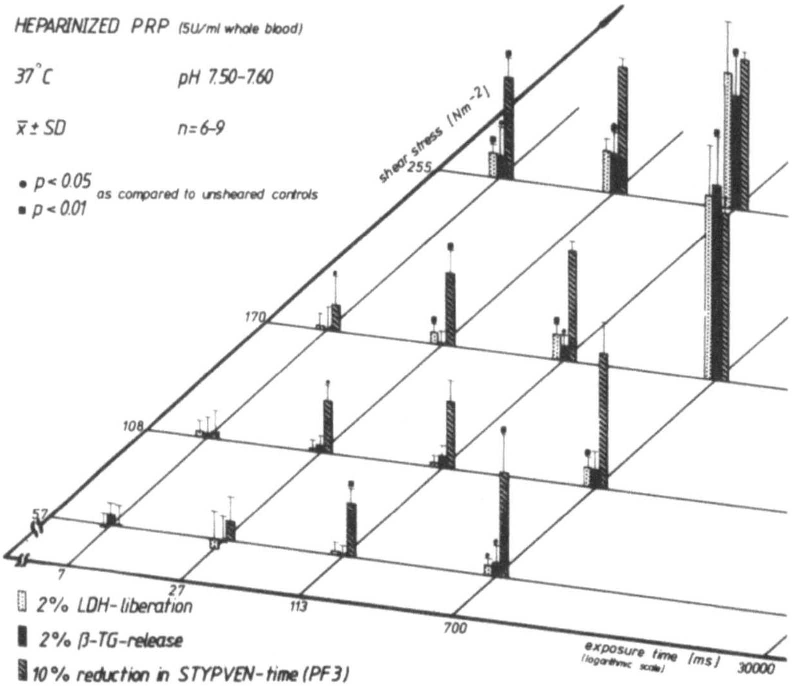

Fig. 2

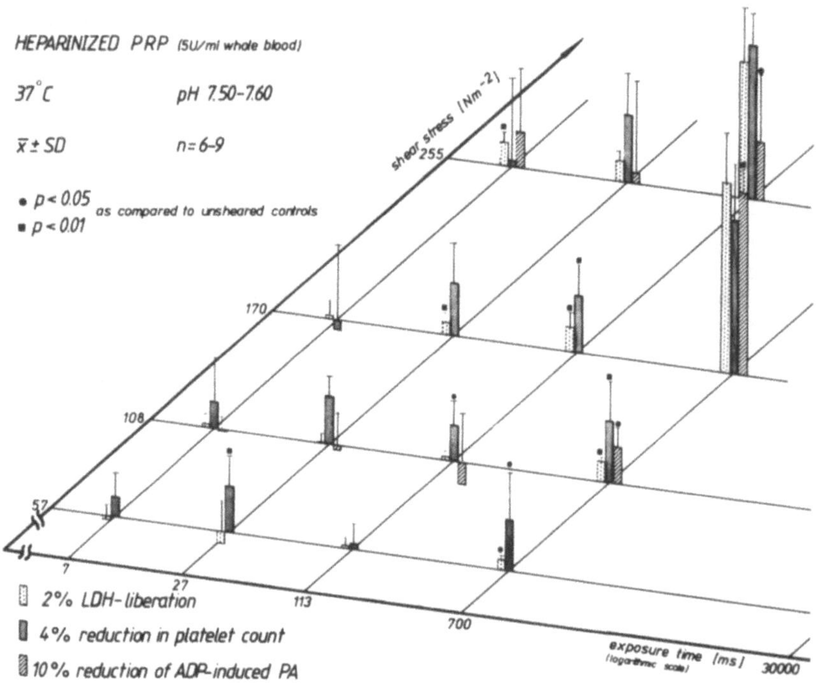

Fig. 3

Figs. 2 and 3. 3-dimensional display of the results obtained after exposing heparinized PRP to high shear for short times. X-axis: exposure time, log. scale; Y-axis: shear stress, linear scale; Z-axis: change in platelet parameter as compared to unsheared control

4. Platelet release reaction (β-TG) and lysis (LDH) need higher levels of shear stress and/or exposure times to become detectable. Interestingly, LDH is set free in our experiments to the same degree as β-TG.

Table 2. % β-TG release after shear exposure

τ \ t_{exp}	7 ms	27 ms	113 ms	700 ms
57 N/m^2	0.9 ± 1.3	0.1 ± 2.7	0.8 ± 1.1	1.4 ± 1.9*
108 N/m^2	0.4 ± 1.8	0.7 ± 1.0	1.1 ± 1.3	1.8 ± 2.1*
170 N/m^2	0.1 ± 2.3	0.1 ± 1.3	1.0 ± 1.2*	19.6 ± 4.9**
255 N/m^2	2.1 ± 3.0*	3.7 ± 2.6**	10.6 ± 4.0**	

x ± S.D.; + $p < 0.05$; ** $p < 0.01$

Table 3. % LDH liberation after shear exposure

τ \ t_{exp}	7 ms	27 ms	113 ms	700 ms
57 N/m^2	-0.4 ± 1.2	-1.0 ± 3.0	0.3 ± 1.0	1.0 ± 0.4*
108 N/m^2	0.4 ± 1.1	0.1 ± 1.1	0.3 ± 0.7	1.9 ± 0.5**
170 N/m^2	0.3 ± 1.4	1.1 ± 1.0*	2.4 ± 1.5**	18.5 ± 5.1**
255 N/m^2	2.2 ± 1.1**	5.0 ± 1.4**	13.3 ± 5.2**	

x ± S.D.; * $p < 0.05$; ** $p < 0.01$

5. Reduction of platelet count is always 2 to 4 times in excess of platelet lysis as calculated from LDH liberation, indicating loss of single platelets into aggregates.

Table 4. % Reduction in platelet count after shear exposure

τ \ t_{exp}	7 ms	27 ms	113 ms	700 ms
57 N/m^2	4 ± 5	9 ± 6**	1 ± 4	10 ± 10*
108 N/m^2	5 ± 9	9 ± 4*	7 ± 5*	12 ± 8**
170 N/m^2	-2 ± 12	10 ± 8*	11 ± 6**	30 ± 11**
255 N/m^2	1 ± 15	13 ± 10*	30 ± 6**	

x ± S.D.; * $p < 0.05$; ** $p < 0.01$

6. Significant impairment of ADP-induced PA is only observed at high exposure times and shear stresses. At medium shear exposure (108 N/m^2, 113 ms) a non significant rise in ADP induced aggregability occurred (Fig. 3).
7. Looking at TEM micrographs of platelets exposed to 170 N/m^2 for 700 ms, one finds dramatic alterations in platelet morphology (Fig. 5). Platelets appear swollen with total destruction of the cytoskeleton, reduction of cell organelles, and decreased cytoplasmic density. Dense tubular system as well as surface connected system are absent in most platelets. The membrane is disrupted at various

Table 5. % Reduction in ADP-induced (5 μM) PA after shear exposure

τ \ t_{exp}	7 ms	27 ms	113 ms	700 ms
108 N/m²	-1 ± 6	-2 ± 16	-11 ± 24	18 ± 10*
170 N/m²				89 ± 11**
255 N/m²	17 ± 32	5 ± 45	28 ± 33*	

x ± S.D.; + $p < 0.05$; ** $p < 0.01$

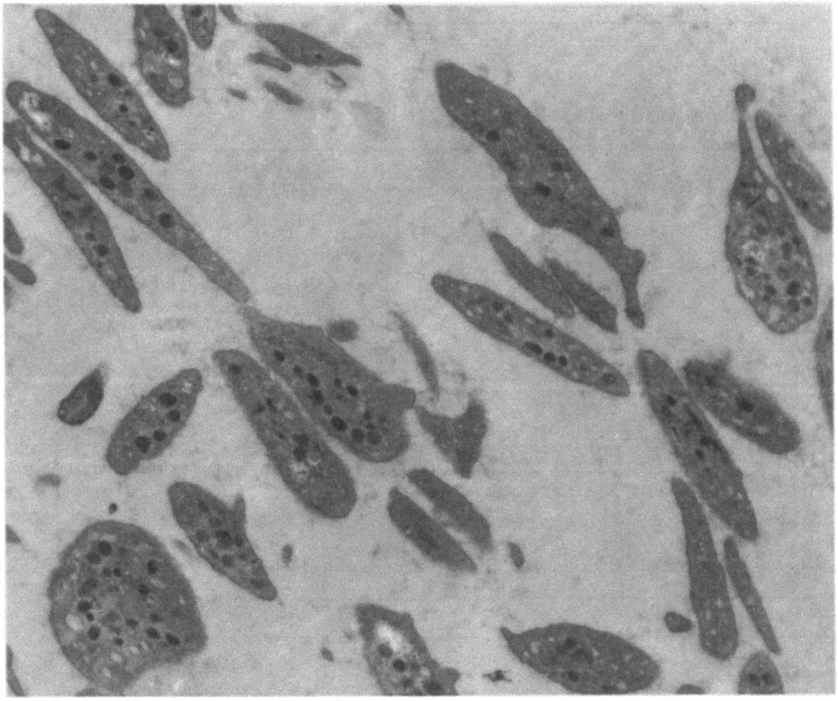

Fig. 4. Transmission electron micrograph (TEMG) showing unsheared ACD-anticoagulated (pH 7.4) disc-shaped platelets with few pseudopodia. (Magnif. 10.750x)

sites of the platelet, forming holes of up to 500 nm (Fig. 6). Free granules and mitochondria indicate total platelet disruption. A number of small aggregates consisting of ca. 10 platelets are found, too.

Following exposure to lower shear stresses (108 N/m^2), 700 ms), platelets exhibit clear signs of activation, in the sense of a shape change from "smooth discs" to "spiny spheres", with their centralized organelles surrounded by microtubules (Fig. 7). Only few platelets show the afore described degenerative changes.

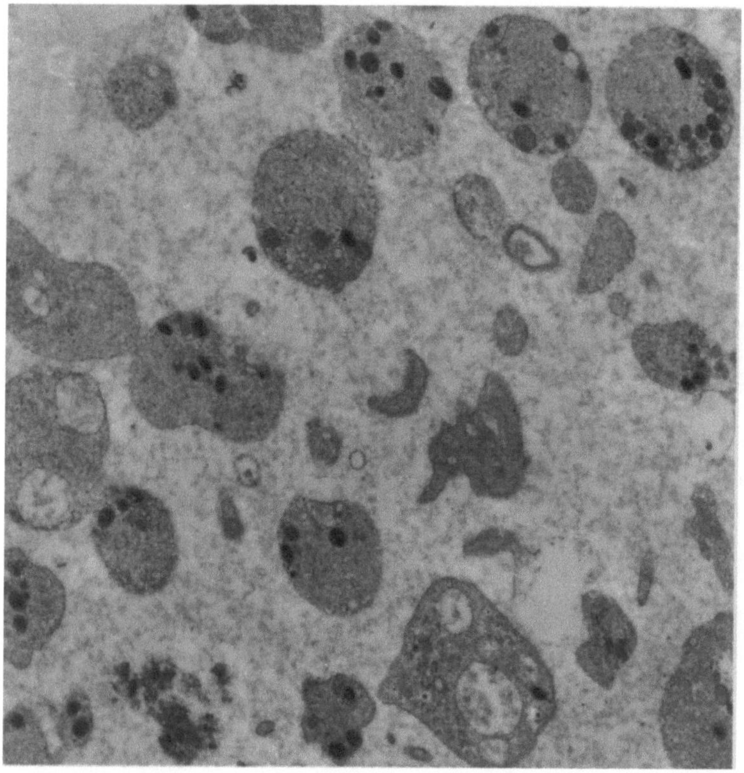

Fig. 5. TEMG from ACD-anticoagulated PRP after 700 ms exposure to 170 n/m². (Magnif. 10.750x)

Discussion

The present work attempts a mapping of various platelet parameters over a range of shear stresses and exposure times, both in the range of pathophysiological interest. Our results can be summarized as follows:

1. Short acting (ms) high shear forces lead to platelet damage, the exposure time being less effective than the magnitude of shear stress.
2. Platelet factor 3 activity is the parameter most sensitive to shear stress.
3. There is no evidence for an active release of platelet constituents beyond passive lysis.
4. ADP-induced aggregation needs rather high shear forces to become suppressed.

Most experiments, published so far, report a so-called shear induced activation of platelets, resulting in a release of ^{14}C-labelled serotonin or β-TG. This is interpreted as a cell-specific active answer to the high shear "stimulus", and is found to occur at lower stress levels than platelet lysis, monitored by LDH liberation [1,2,7,16].

The data presented here, provide no evidence whatsoever to support this notion. The discrepancy with previous investigations could be due to the long time platelets are exposed to high shear in the quoted studies. There is reason to believe that the contents (e.g. serotonin, ADP) of a few platelets, lysed already at lower shear stresses, might

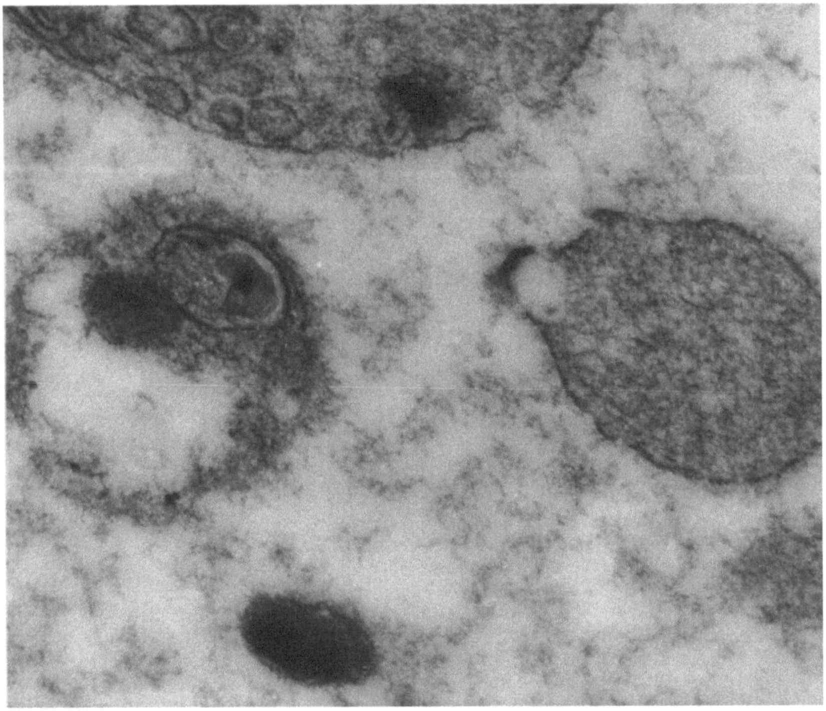

Fig. 6. TEMG from the same sample as Fig. 5, showing membrane defects and membrane-coated free granules. (Magnif. 60.000x)

indirectly stimulate undamaged cells to release reaction, leading via a positive feed back during minutes to a higher liberation of released β-TG than LDH. The rise in PF 3 activity at significantly lower stresses seems to contradict this explanation. However, Zwaal and coworkers provided evidence, that lysis rates below 1%, escaping definite confirmation by our conventional enzymatic methods, lead to a significant drop in RVV-activated plotting time [22]. The shape change of platelets at "sublytic" shear stress levels, documented by TEM, could be caused by a few lysed platelets. Shape change in contrast to release can well happen within the few seconds until platelets are processed.

Nevertheless, the active response of platelets to the liberated contents of other blood elements may be of pathophysiological relevance, possibly for the initiation, but even more likely for the progression and complication of existing lesions, irrespective of their cause. For example, in stenoses hydrodynamic forces, shown to lyse platelets can by expected. In an arterial stenosis of ca. 90% cross-sectional reduction over a length of 10 mm, leading to a pressure drop of 40 mmHg, one can estimate that platelets travelling in 5 μm distance from the wall, encounter a shear stress of 145 N/m^2 for about 50 ms. Roughly 1% of the platelets in this layer will lyse, and will therefore deliver both procoagulant activity and mediators (such as serotonin, platelet derived growth factor), that have the potential to damage endothelium or to stimulate smooth muscle cell proliferation. Laminar undisturbed tube flow will eventually dilute and wash away their thrombogenic contents. More importantly, a stenosis is usually followed by a poststе-

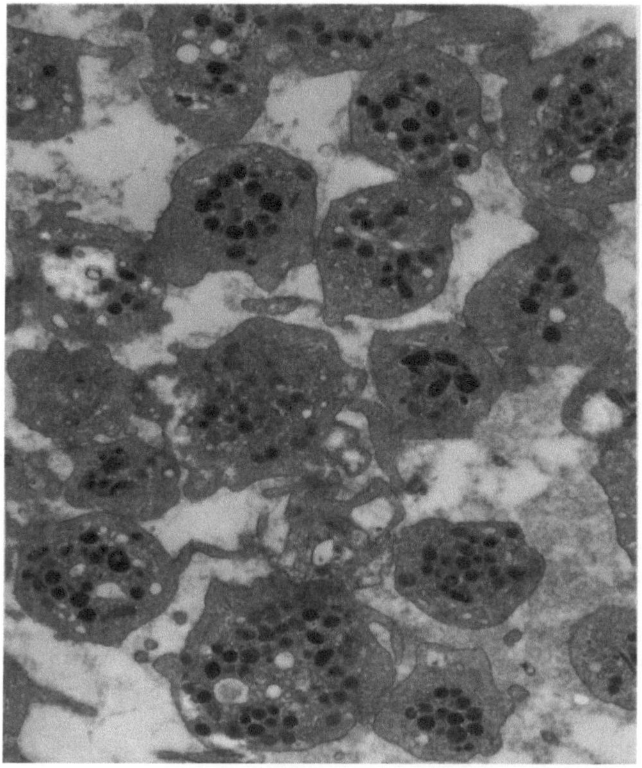

Fig. 7. TEMG showing platelets after 700 ms exposure to 108 n/m². All platelets have undergone shape change. (Magnif. 10.750x)

notic dilatation with extensive recirculation. Such recirculation zones are preferentially fed with fluid elements from wall-adjacent layers i.e. high shear regions, especially in pulsating flow when the extension of the eddy changes periodically. Comparatively low shear rates, and most important, limited convective mass exchange provide hydrodynamic conditions akin to that in a reaction tube or aggregometer in the laboratory. The contents of lysed platelets as well as red cells [19] may activate unaltered platelets, which find favourable conditions to aggregate and release their granular contents, thus augmenting the thrombotic potential of the blood via a positive feed back mechanism.

However, even maximally activated platelets do not seem to adhere to walls, no matter how thrombogenic they are, unless hydrodynamic forces provide transport to the wall. As shown in the companion paper (Blasberg et al.), in undisturbed tube flow, even ADP-activated platelets do not adhere. Only flow vectors directed towards the wall, as they are regularly found at the reattachment point, lead to platelet adhesion, the extent of which is modified by their degree of biological activation [3].

This described concept of thrombus formation in high flow regimens has been substantiated experimentally in its 3 components (Fig. 8):

1. High wall shear stresses in a stenosis by damaging platelets and red cells constitute a potent thrombogenic stimulus.
2. The effects of this physical stimulation become amplified enzymatically in the recirculation zone.

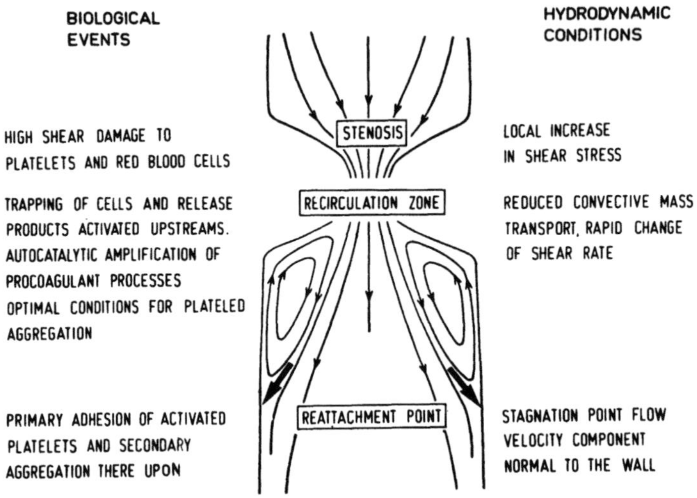

Fig. 8. Hypothetical scheme depicting the sequence of rheological events leading ultimately to white thrombus formation

3. Flow directed towards the wall leads to manifestation of the thrombotic deposition at a site, different from where it is triggered off.

While all these 3 steps require different hydrodynamic boundary conditions, they are, nevertheless, provided in arterial stenoses. At the predilection sites of atherosclerosis e.g. branchings, blood cells indeed can pass an identical sequence of flow fields (Karino et al., this volume).

Modelling a stenosis we could experimentally prove that in a poststenotic recirculation zone the coagulation potential (PF 3) is raised indeed locally, as compared to blood up or downstream.

References

1. Anderson GH, Hellums JD, Moake JL, Alfrey CP (1978) Platelet lysis and aggregation in shear fields. Blood Cells 4: 499-507
2. Anderson GH, Hellum JD, Moake JL, Alfrey CP (1978) Platelet response to shear stress: changes in serotonin uptake, serotonin release, and ADP induced aggregation. Thromb Res 13: 1039-1047
3. Baldauf W, Wurzinger LJ, Kinder J (1978) The role of stagnation point flow in the formation of platelet thrombi on glass surfaces in tubes with various geometry. Path Res Pract 163: 9-33
4. Bernstein EF, Marzec U, Johnston GG (1977) Structural correlates of platelet functional damage by physical forces. Trans Am Soc Artif Intern Organs 23: 617-624
5. Blasberg P, Wurzinger LJ, Mussler K, Myrenne H, Schmid-Schönbein H (1981) A platelet aggregometer with automatic data processing. Thromb Haemostas 46: 132
6. Cella G, Zahavi J, de Haas A, Kakkar VV (1979) β-Thromboglobulin, platelet production time and platelet function in vascular disease. Br J Hemat 43: 127-136
7. Colantuoni G, Hellums JD, Moake JL, Alfrey CP (1977) The response of human platelets to shear stress at short exposure times. Trans Am Soc Artif Intern Organs 23: 626-630
8. Hardisty RM, Hutton RA (1966) Platelet aggregation and the availability of platelet factor 3. Br J Hemat 12: 764-776

9. Heuser G, Opitz R (1980) A Couette viscometer for short time shearing of blood. Biorheology 17: 17-24
10. Ludlam CA, Moore S, Bolton AE, Pepper DS, Cash JD (1975) The release of a human platelet specific protein measured by a radioimmunoassay. Thromb Res 6: 543-551
11. Morgenstern E (1980) Ultracytochemistry of human blood platelets. Progr Histochem 12 (4)
12. Ranstack JM, Zuckerman L, Mockros LF (1979) Shear-induced activation of platelets. J Biomech 12: 113-125
13. Rieger H, Wurzinger LJ, Schmid-Schönbein H (1977) Influence of temperature on shear induced platelet aggregation in vitro. Klin Wschr 55: 121-130
14. Rokitanski Cv (1841) Handbuch der pathologischen Anatomie. Braumüller und Seidel, Wien
15. Ross R, Glomset JA, Kariya B, Harker LA (1974) A platelet-dependent serum factor that stimulates the proliferation of arterial smooth muscle cells in vitro. Proc Nat Acad Sci (Wash) 71: 1207-1210
16. Stevens DE, Joist JH, Sutera SP (1980) Role of platelet prostaglandin synthesis in shear-induced platelet alterations. Blood 56: 753-758
17. Williams AR, Chater BV (1980) Mammalian platelet damage in vitro by an ultrasonic therapeutic device. Arch Oral Biol 25: 175-179
18. Wurzinger LJ, Blasberg P, Schmid-Schönbein H (1980) Hemolysis and in vitro thrombus formation. In: Stoltz JF, Drouin P (eds) Hemorheology and diseases. Doin, Paris pp 529-533
19. Wurzinger LJ, Blasberg P (1981) The influence of anticoagulants upon the results of platelet function tests: aggregation and release reaction. Thromb Haemostas 46: 419
20. Yu SK, Latour JG, Marchandise B, Bois M (1978) Shear stress-induced changes in platelet reactivity. Thrombos Haemostas 40: 551-560
21. Zahn FW (1875) Untersuchungen über Thrombose: Bildung der Thromben. Virchows Arch path Anat 62: 81-88
22. Zwaal RFA, Rosing J, Tans G, Bevers EM, Hemker HC (1980) Topological and kinetic aspects of phospholipids in blood coagulation. In: Mann KG, Taylor FB (eds) The regulation of coagulation. Elsevier, North Holland, New York Amsterdam Oxford, pp 95-112

Discussion

C. SCHWARTZ: I wonder, can you comment on the nature of the mechanism that is responsible for the reorientation of the platelet organelles in the system? Also, why in the shear state does there appear to be primarily a loss of the cytoplasmic processes? Thirdly, have you any comments on whether platelets derived from hyperlipidemic serum, perhaps with increased membrane lipid content, would behave the same or differently in different shear states?

WURZINGER: To answer your first question, in light of our morphological observations, we think that the membrane becomes defective, at least for a short time, and that surrounding fluid leaks in. As we all know, the calcium level in citrated and even more in heparinized platelet-rich plasma is much higher than in the platelet's cytoplasm. This leads among other things to a disintegration of the microtubular cytoskeleton. Of course, it seems conceivable that the mechanical strain itself might damage the cytoskeleton, but as I already mentioned, the influx of plasma with its high calcium content suffices to explain the disintegration of the cytoskeleton. The pronounced reduction in cytoplasmic density suggests quite a considerable influx of plasma into the mechanically damaged platelets. In the phase contrast microscope, we observe ballooned platelets, with their granules moving in the cytoplasma with Brownian motion, which we take as a strong argument for the total breakdown of the internal structure of at least some platelets. With regard to your last question, I am afraid I cannot answer it.

BLACKSHEAR: We had a lot easier time lysing red blood cells in an elongational stress field than we did in a shear stress field. Others, in studying the rupture of oil droplets in water, found it much easier to elongate and rupture oil drops in an elongational field than in a shear stress field. I thus think it is quite possible in your experiment in the converging/diverging flow system that it was the acceleration of the platelet into the orifice that may have done the damage rather than the shear stress downstream of the orifice. This might be worth looking into because it has special relevance to the problem of turbulence. There exist small elongational fields in a turbulent field that could be much more damaging than the shear stress.

WURZINGER: If you are referring to the experiments with the stenosis model presented at the end of my paper, you are right; we cannot exclude that the cells are damaged when they are accelerated entering the tube. With regard to the Couette apparatus, the clear dependence of the observed effects on exposure time leaves no doubt that these effects are produced within the high shear region.

ANONYMOUS: I estimate the degree of your stenosis to be 96 percent. In the body this would be flow reducing, so the flow rate would be less. Would you care to comment?

WURZINGER: We used an extreme stenosis, which can be considered representative only for the late stages of the disease we are talking about. On the other hand, the experiment lasts for minutes and the organism has to cope with, admittedly lower shear stresses, but for decades.

STEHBENS: Dr. Wurzinger, in your electron micrographs the first one did in fact show some changes and in the second obviously they remain spherical with a few dendritic processes. Did you in your model have a control system where you actually circulated the platelet-rich plasma without a stenosis?

WURZINGER: We of course had controls examined in the electron microscope. The first electron micrograph on which you are commenting, with the fairly smooth disc-shaped platelets, was taken from a sample having passed the apparatus without exposure to high shear. They were pumped through the Couette slit with the inner cylinder resting. We also have taken controls from the platelet reservoir, before passage of the apparatus, and we have found no difference from the ones shown.

Microrheology of Thrombocyte Deposition: Effect of Stimulation, Flow Direction, and Red Cells[1]

P. Blasberg, L. J. Wurzinger, and H. Schmid-Schönbein

A. Introduction

The vessel wall, the vessel content and its flow behaviour form a highly complex regulative circuit, which is until now understood only in a few of its manifold interactions. The flow forces and the flow patterns for example are able to affect the formed blood elements as well as the vessel wall. The geometry and the biochemical metabolism of the vessel wall will in turn influence blood flow and properties of blood cells. All components that interact may be considered to be involved in atherogenesis. Amongst the formed elements of the blood, the platelets are suspected to play a paramount role in the development of arterial thrombosis and the early stages of arteriosclerosis.

In various studies in vitro (Feuerstein et al. 1975) and in vivo (Begent and Born 1970; Bergqvist and Arfors 1977) as well as in perfusion experiments in arteries denuded of endothelium (Antonini et al. 1978; Turitto and Baumgartner 1979) it has been shown that platelet deposition is augmented with increasing shear rate and hematocrit (i.e. red cell volume fraction).

"Enhanced platelet diffusivity in flow" has been proposed as explanation for the positive correlation between flow (or shear) rate and platelet deposition in flow chamber experiments. This presently popular hypothesis assumes that stirring motions of red cells in rapidly flowing blood occur which would move platelets to the site of deposition at a higher rate than expected from Brownian motion. In support of this assumption, experiments by Goldsmith et al. (1975) are often cited in which the trajectories of latex particles (as models of platelets) in a suspension of red cell ghosts (as models of intact erythrocytes) have been recorded and the deviations from straight trajectories were found larger in latex particles than in red cells. These experiments, however, were only executed in creeping flow (wall shear rate < 20/s) and have neither been confirmed by experiments with platelets nor in experiments at the high shear rates (> 100/s), where enhanced platelet deposition is usually found. At these high shear stresses, mammalian red cells - in striking contrast to all other known particles, but in accordance to the behaviour of fluid droplets - do not rotate in shear flow, but rather are deformed into prolate ellipsoids with their membrane rotating around the fluid cytoplasm, and they are in a stationary orientation with the major axis of the ellipsoid oriented roughly in parallel to the plane of shear (Fischer and H. Schmid-Schönbein 1977). At very low shear rates the red cells are not deformed and behave more like ordinary particles.

1 Supported in parts by Fritz Thyssen Foundation, Cologne

Fluid Dynamics as a Localizing Factor for Atherosclerosis
Editors: G. Schettler et al.

Before the experimental proof of an enhancement of platelet motion across the planes of shear by red cell rotation is established at high rates of shear, we remain sceptical of this explanation for the enhancement of platelet deposition in flow. In addition, there are theoretical reservations, since the observations of particles in the experiments of Turitto and Baumgartner (1979) and of Leonard (1972), the distances travelled by the particles are very many times larger than the average spacing of the platelets in the blood, and larger than the mean free path between platelet-platelet collisions. The average spacing of platelets in blood is 15 µm, calculated by assuming that $3x10^5$ are suspended in 1 µl blood or 0.55 µl plasma.

It is known from the kinetic theory of gases that a gas can be considered as a continuum (for which the analysis of diffusion driven by a concentration gradient can be applied) if the dimensions for the problem are large compared with the mean free path. For developing mural thrombi, about 10 - 100 µm long, this condition is not satisfied. Over the length of the thrombus the main flow does not have enough time to "learn" of the platelet deposition at the wall to drive more platelets towards the wall by diffusion down the concentration gradient. We can also estimate the relative importance of platelet-platelet or platelet-thrombus collisions caused by convective movement in a velocity gradient compared with Brownian motion. This ratio is expressed by the Smoluchowski number:

$$S_M = \frac{\text{frequency of encounter by velocity gradient}}{\text{frequency of encounter by Brownian motion}} = \frac{4\dot{\gamma}R^2}{3\pi D}$$

where R = particle radius, $\dot{\gamma}$ = shear rate, D = diffusion coefficient.

This number was calculated by Richardson (1973) using data from the experiments of Begent and Born to be nearly 10^4. Even if we assume the diffusivity is increased 10^2 by augmentation in shear flow, the number would be ca. 10^2, which would mean that the diffusion effects are unimportant compared with those due to the main convective motion. As an alternate explanation for the well established positive effect of hematocrit, shear rate and/or shear stress we have previously postulated that the flow itself has effects on the biological and rheological behaviour of platelets, "activated" by fluid-dynamic forces. It has been previously shown that high fluid dynamic forces can alter their microrheological and adhesive properties (H. Schmid-Schönbein et al. 1981). Using a new flow apparatus, this concept was further pursued in the experiments described in the companion paper by Wurzinger et al. The present study deals with the influence of various flow conditions on the microrheological behaviour of normal and activated platelets in tube flow, focussing on the differences between flow parallel to a wall, normal to a wall. Furthermore, interactions between red cells and platelets in the immediate vicinity of a wall are assessed. Our data further support the notion that there are very significant flow dependent variations in platelet microrheology which have to be included in any analysis of macroscopic thrombotic events brought about by platelets.

B. Material and Methods

I.1. Glass Models of Arterial Branchings and Bends. To study the preferential sites for the deposition of platelets in disturbed flow as it occurs in branched and curved vessels glass models were used. These studies were carried out in collaboration with Kratzer, Baldauf and Kinder in the laboratory of Müller-Mohnssen and are described in detail in the dissertation of Wurzinger (1979). The models had the following dimensions: the branched models had inner diameter of 2.6, 3.0 and 3.4 mm in the main tube and 1.5 to 2 mm in the branch; the mouth of the side branch was blown out of the wall of the main tube and had the shape of a truncated cone to which the actual side branch, a straight tube was fixed. The branch angle was 90°. Details of the flow in these models of coronary arteries are given by Müller-Mohnssen (1971). Tubes curved to 180° with an i.d. of 3.4 mm and a radius of curvature of 6.8 mm were also used.

I.2. Platelet Suspensions. Bovine platelet rich plasma (PRP) was chosen because bovine platelets - in contrast to human platelets - do not adhere to glass unless chemically stimulated (Marzec et al. 1975). ADP with different concentrations in the µmolar range was added to the bovine PRP, which allowed to produce controlled states of platelet "activation". Platelet concentrations were adjusted to values between 15,000 and 200,000 per µl. The majority of the experiments were carried out with a platelet count of 50,000/µmol ADP/l.

I.3. Microscopical Observation. The movement and deposition of platelets was observed by reflective darkfield microscopy (Baldauf et al. 1978).

II.1. Models for Undisturbed Tube Flow. To investigate details of the flow behaviour of platelets in undisturbed laminar tube flow polyvinyl chloride (PVC) and glass capillaries with i.d. between 100 and 300 µm were used. The capillaries were pulled from PVC tubing or hematocrit glass capillaries over a flame and checked microscopically for constant diameter. The capillaries were mounted on a motor driven microscopical stage allowing the motion of the capillaries opposite to the flow direction at different velocities ("travelling microscope"). Moving the perfused vessels at equal but opposite direction as suspended particles at a certain streamline this technique allows the observation of cells undergoing shear flow at fictitious standstill. While examining certain streamlines the number of platelets per unit volume was determined using a manual counting technique based on closed circuit TV tape-recordings of the flow experiments (v.i.). In addition, motions of particles perpendicular to the streamlines and collisions with the wall were observed. The tubes were connected to a constant rate syringe which was set to yield wall shear rates between 50 and 100/s in the perfusate.

II.2. Perfusion Fluids. Human blood was used and the platelets were separated by centrifugation and stained with the fluorescent dye acridin orange (AO) (final concentration 20 or 80 µmol/l). AO is taken up by the platelets rapidly and bound to polyanions in such a fashion that molecules isolated from one another, show green orthochromatic fluorescence whereas when bound to neighboring sites on polyanion molecules and aggregated with one another, exhibit red metachromatic fluorescence (Wolf and Aronson 1961). Platelets stained with AO show a green cytoplasmic and a red granular fluorescence. After staining for 5 min the platelets were fixed with glutaraldehyde (final concentration 0.25%) for 5 min. In light of the great susceptibility of human platelets to undergo shape change after removal from the circulation, special care was taken to maintain their native discoid configuration. All steps from withdrawal (ACD-anticoagulation) to fixation were carried out keeping strict isothermia at 37°C. The fixative was removed from the plasma by repeated (3x) washings and the platelets then resuspended in autologous plasma. Concentrated red cells were then added to yield the desired hematocrits, thus ensuring a constant platelet concentration per free plasma volume despite the presence of red cells. In addition, a small number of experiments were also carried out with unfixed but stained platelets to investigate details of the deposition onto the walls.

II.3. Microscopical Observation. Epifluorescence microscopy was employed using a 200 W high-pressure mercury lamp for UV and blue light excitation. A TV-camera with a target extremely sensitive to red light and a video tape recorder were used for documentation. In most of the experiments streamlines with a distance of 4 µm from the wall were observed. A special particle counting technique in flow was developed. For the applied wall shear rates the backward velocity of the microscopical stage was chosen in such a way that the streamlines at a distance of 4 µm from the tube wall (as calculated from a parabolic velocity profile in accordance with the data of Bennet 1968) could be observed because here the flowing platelets appeared to be stationary. Platelets flowing on those streamlines were counted subsequently by repeated playback of the recorded videotapes stopping the tape recorder at defined time intervals. These time intervals correspond to different path lengths travelled by the particles at different flow rates.

C. Results

I. Bovine Platelet Deposition in Branched Tubes Perfused with Bovine PRP: Effect of Flow and Activation

I.1. Flow Effects. Two predilection sites for polycellular platelet deposits in the main branch were found:

1. at the distal rim of the side branch oreface (flow divider),
2. at the wall opposite and downstream from the branching point;

deposits at the flow divider occur irrespective of the flow rate and the fractional flow rate into the branch. When the flow rate into the branch was zero, there was a small eddy at the branch entrance which was sufficient to deposit platelets on this site. Deposits also developed when highly diluted PRP (15,000 platelets per µl) was used. The deposits usually appeared as stripes or tailed spots oriented in the direction of the streamlines close to the wall (Fig. 1).

Platelet deposition on the downstream wall opposite the orifice of the side branch occurred only when flow separation and instationary flow with periodic oscillations in the main tube downstream from the branching point was induced. The appearance of these flow instabilities in steady perfusion only depended on the fractional flow rate through the side branch within a wide range of flow rates through the main tube. A simple, stationary recirculation zone and its reattachment point were not sufficient to produce platelet deposition. In pulsatile flow, the flow instabilities as well as the downstream depositions occurred at lower fractional flow rate through the branch. Using the method of Kinder and Kratzer (1975) for flow visualization by pulsed laser light it could be shown (Fig. 2), that the area where platelets were deposited coincided with the region where streamlines were directed towards the wall. The extent of deposits was found to be positively correlated to the platelet concentration as well as to the ADP concentration (details see Wurzinger 1979; Baldauf et al. 1978). For technical reasons (rough surface, opaque glue) the deposition patterns in the side branch could not be studied in detail.

I.2. Effects of Activation. Using a wide range of Reynolds numbers in the branched and curved glass models of arteries, there was no platelet

Fig. 1. Platelet deposits in the region of the distal rim of the side branch origin (main flow from left to right). Stripes of deposits diverge from the stagnation point on the edge of the origin of the side branch following the streamlines of the secondary flow close to the wall

deposition in the straight inflow and outflow tubes, neither with native nor with stimulated platelets. Likewise, no deposition was found near the branching or bends when unstimulated bovine platelets were used. However, when stimulated with ADP, the platelets deposited rapidly at characteristic sites, namely at the flow divider and at the attachment points of laminar flow behind a recirculation zone.

II. Platelet Deposits in Bent Glass Tubes (Fig. 3)

As in the above experiments, there was no significant platelet deposition in bends unless they were previously stimulated by ADP. Subsequent to chemical "activation" platelet deposits appear at two sites within curved tubes:

1. on the outer wall (i.e. the wall with the greater radius of curvature) when the Re-number was greater 530 (platelet count: 100,000/ μl; $c_{ADP} = 5\times10^{-6}$ m). The polycellular deposits were oval in shape with their long axis aligned parallel to the streamlines close to the wall.
2. on the inner wall for Re > 720. Thereby backward movement of platelets was caused along the inner wall, indicating an area where the flow has separated from the wall. Under such flow conditions deposits could be found on the inner wall of the bend where the separated flow reattaches to the wall forming a stagnation point between the reattaching and the recirculating streamlines, i.e. where forward and backward flowing particles are separated.

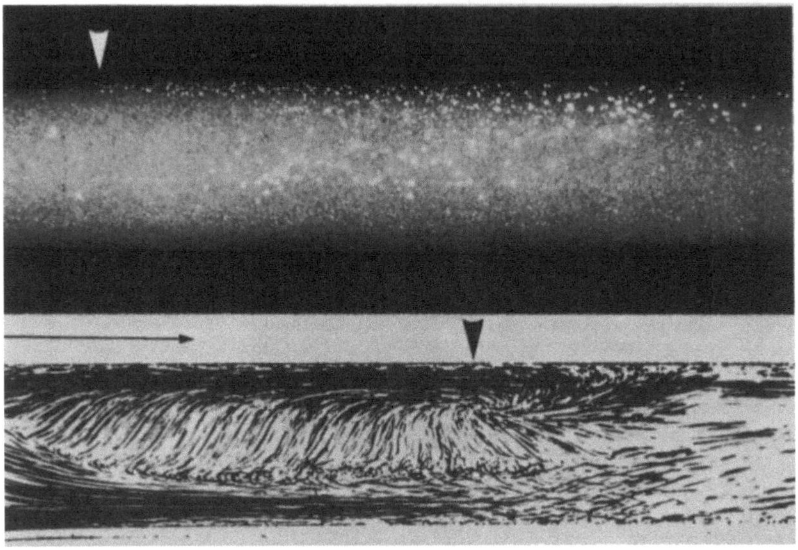

Fig. 2. *Top:* Platelet deposits on the wall of the main tube opposite the origin of the side branch and 9 to 18 mm downstream from the branching point. *Bottom:* Streamlines of flow leading to deposits as shown in the top picture, visualized by laser ultramicroscope illumination of a thin layer of fluid in a plane containing both axes of the branched tube (picture by M. Kratzer). Tips indicate in both pictures a distance of 9 mm downstream from the branching point

III. Platelet Adhesion in Straight Tubes Perfused with AO-Labelled Human PRP

In studies comparing PVC and glass capillaries, these were perfused with AO-stained human PRP with unfixed platelets and observed microscopically. Due to the great susceptibility of human platelets to be activated by unspecific stimuli such as they occur in preparing PRP, we have to assume that at least some of the platelets had undergone "viscous", i.e. sticky metamorphosis. In keeping with the observation with bovine platelets, no depositions developed in the glass tubes. In contrast, in PVC tubes wall adherent platelets were often found, despite the fact that PVC is considered to be less thrombogenic. This deposition was clearly related to wall roughness: wall protrusions up to 7 µm of the PVC tubes used were found whereas the glass tubes did not show microscopically detectable roughness. The platelet deposits in the PVC tubes were always localized a few microns downstreams from those wall protrusions, which generated a local flow disturbance with subsequent reattachment of streamlines to the wall. If one larger platelet deposit had formed it represented itself as an additional flow disturbance and a second deposition was regularly found downstreams.

IV. Platelet Enrichment in the Vicinity of the Wall in the Presence of Red Cells

The addition of red cells to PRP showed a striking effect on platelet density near the wall, which depended on hematocrit and flow rate. To quantify this effect glass capillaries were perfused with AO-labelled and fixed platelets at different hematocrits and flow rates.

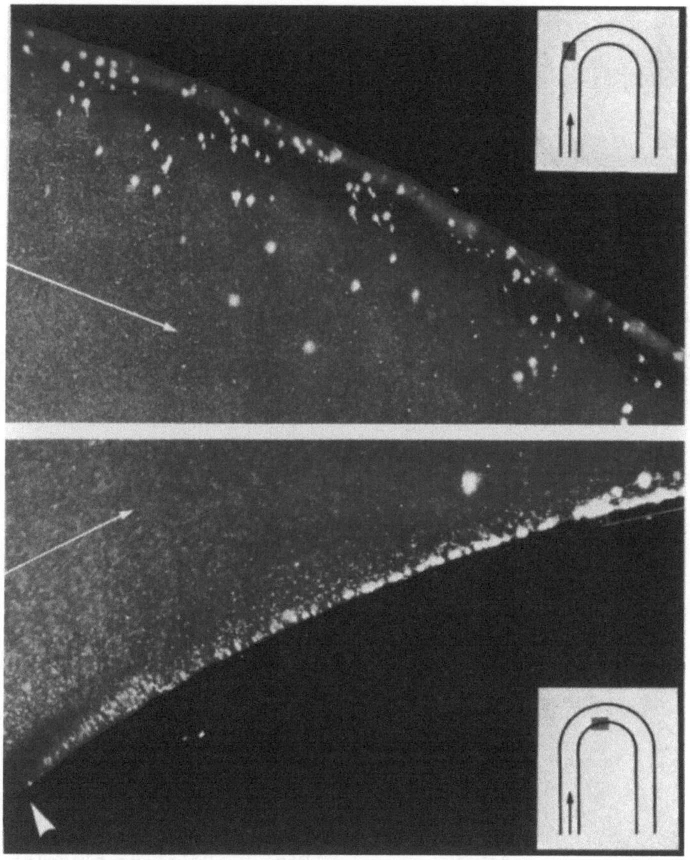

Fig. 3. *Top:* Platelet deposits on the outer wall of the bend. *Bottom:* Platelet deposits on the inner wall, found 1 mm downstream from the point of flow separation (tip) (direction of main flow as indicated by arrows; inserts give the positions of the photographed areas)

When only PRP was perfused, i.e. in the absence of red cells only a few platelets were observed in the vicinity of the wall. Some platelets were even closer to the wall than 4 µm, but at no instance platelets could be observed, which achieved contact with the wall. Rising shear rates depleted the marginal layer completely of platelets. Slight lateral movements were only found when platelets rotated in flow (details will be given elsewhere).

There was a drastic change when red cells were added to the PRP, even at low concentrations. Rising hematocrit strongly elevated the concentration of platelets in the marginal layer. As can be seen (Fig. 4), the platelet density increased significantly with rising hematocrit, with a 12fold increase at hematocrit 0.33, with even higher platelet crowding at 0.45 (not shown). As shown in Fig. 4, this marginal concentration of platelets seems to be nearly unaffected by the applied shear rate between 150/s and 900/s. However, if one calculates the total flow of platelets (or platelet discharge) through the observed layer by multiplying the platelet density by the absolute velocity in the layer, this value further increases (Fig. 5). Therefore, by increasing the hematocrit to 33% and the shear rate to 910/s, the total platelet flux can be increased by a factor more than 70 compared to that found in PRP at 150/s.

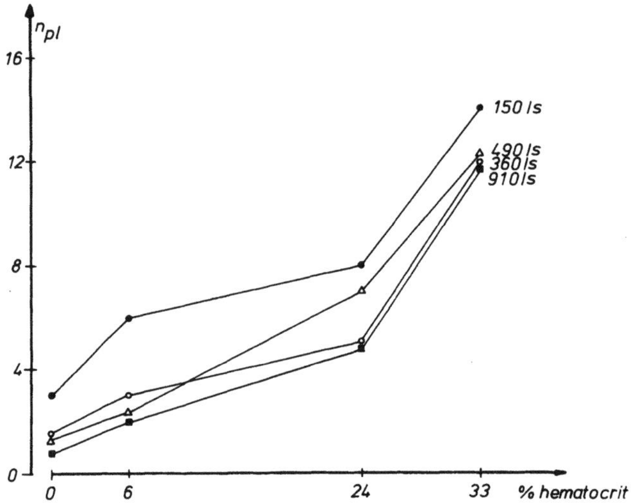

Fig. 4. Relative platelet numbers observed at a distance of 4 µm from the wall at given time intervals for different wall shear rates and hematocrits (platelet concentration: 250,000 per µl free plasma)

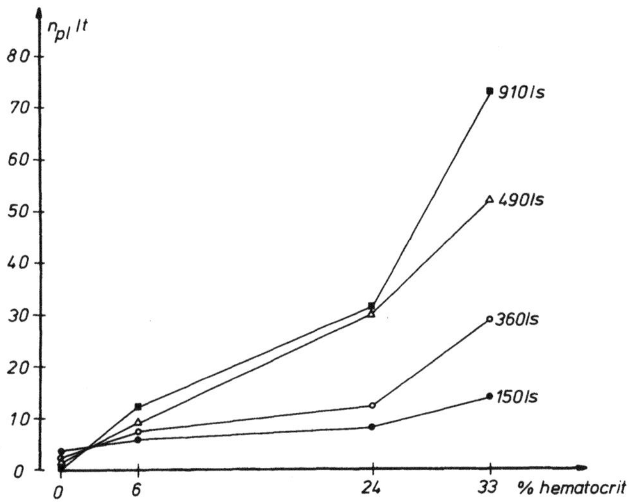

Fig. 5. Relative platelet numbers observed at a distance of 4 µm from the wall of a 150 µm (i.d.) capillary per unit observation time (platelet concentration 250,000 per µl free plasma)

A technical problem arose when platelets were observed for a given hematocrit and rising flow rates due to the limits of observation times. Because of the poor time resolution of the used video system and the different flow velocities of the observed streamlines at different wall shear rates, it was often difficult to compare the platelet concentrations in these series. However, it could be seen clearly that rising shear rates have a strong effect on platelet movement perpendicular to the streamlines. Secondary motions of the platelets were seen as erratic back and forth motions and movements in and out of focus. We arbitrarily assumed wall contact when the platelets appeared to move backward with the same velocity as the travelling wall. The red cells generate a local flow disturbance that bounce platelets against the wall and away from it. Even though the frequency of these events is difficult to quantify, we clearly say that the frequency of these phenomena rose conspicously with increasing shear rates but was less affected by hematocrit.

D. Discussion

In experiments with PRP-perfused glass models of arteries it could be demonstrated that platelets have to be brought by convective forces to the wall as a prerequisite for sticking to it. Convection occurs in a stagnation point flow, which naturally happens at every flow divider, but also in all forms of secondary flow associated with the formation of areas with flow separation, stagnation point flow and the reattachment point, which is known to favour the deposition of platelets (Karino, this volume).

In the presence of red cells additional flow components towards the vessel wall are present also in undisturbed overall laminar tube flow. The red cells seem to act by two hydrodynamic mechanisms:

1. to enhance platelet concentration near the vessel wall (their target site under both physiological and pathological conditions) and
2. at high shear stresses, they produce erratic motions of the platelets crowded in the marginal layer. These erratic motions include a frequent encounter with the vessel wall, which in turn favours permanent platelet deposition at hydrodynamically distinct sites. Such sites can be "distinct" by microscopic roughness or even by the presence of already adherent platelets. The frequency and strength of such "physiological" platelet-wall contact are probably lower than at sites where the blood stream hits the vessel wall directly; but this remains to be determined quantitatively.

The results of the microrheological studies reported here allow a reevaluation of the data and concepts presented by previous authors who found that the platelet deposition rate depends on flow rate (or wall shear rate) and the presence of red cells. The deposition rate in flow reflects the net effect of platelet flux towards the wall, the nature of the cohesive forces between platelets and wall and the resistance against the hydrodynamic forces that detach and/or remove the platelets from the sites of deposition. First we would like to stress that among all blood borne cells, the platelets can be singled out by their tendency to stick to the wall which depends on a number of factors:

1. they occur at high concentration in whole blood, but especially in the wall adjacent to layers
2. they are small and have practically the same density than the plasma
3. they can be stimulated quickly and in undergoing "viscous" or sticky metamorphosis adhere to the wall and to other platelets and
4. they exert pronounced biochemical and enzymatic activities which are known to affect the endothelial cells, other platelets and the coagulation system (see paper Wurzinger).

These characteristic features of the platelet supply an alternate hypothesis for the flow and red cell dependence of platelet deposition. We feel, as indicated in the introduction, that the assumption of enhanced platelet diffusivity (Keller 1981) is no longer necessary. The marginal accumulation of platelets and 70fold increase in the platelet convection is by order of magnitudes higher than any conceivable diffusion process. The extremely enhanced platelet-platelet collisions and most of all their significantly augmented convective flux is a much more likely explanation for the shear and red cell enhanced platelet availability at the site of thrombotic deposition. We also have strong reservation about the interpretation of the experiments that were cited to support the notion of enhanced platelet diffusivity. The model experiments of Goldsmith who studied the trajectories of latex particles when admixed to red cell ghosts were performed at such low shear rates (< 20/s) that they cannot be accepted as the explanation of rheological phenomena at high shear rates in whole blood. At these low shear rates, red cell axial migration and platelet marginal accumulation cannot be expected to occur. Keller's assumption that red cells rotate in flow and thereby force platelets towards the wall has always been in conflict with established knowledge about the rheological behaviour of red blood cells which are aligned and oriented in flow and undergo continuous tank treading of the membrane (Fischer et al. 1978). It remains to be seen whether this fluid drop like behaviour of the red cell is responsible for the observed accumulation of platelets and for the erratic platelet motions at high hematocrit and at high shear rates. Our observations suggest, however, that this influence is at best a long range effect since we find a very few red cells near the layers in the wall but nevertheless erratic platelet motions.

Mammalian red cells and white cells differ significantly in these respects: while red cells are far more frequent, they are rarified near the wall due to their capacity to migrate axially. As a preliminary explanation of the results obtained, we would like to propose that the mechanism of axial migration is one of the principal causes of the dramatic enrichment of platelets near the wall. By a mechanism that awaits elucidation, the red cells migrating toward the vessel axis appear to displace the platelets towards the wall.

Neither red cells nor white cells appear to be able to stick to the wall. While red cells can be accumulated in areas of low shear, they are washed away as soon as the flow - and thence the shear stresses are elevated. White cells do not appear to be adherent to subendothelial structures - but they are capable of sticking to the apparently intact endothelium (G. Schmid-Schönbein et al. 1980) although only under very specific conditions such as they occur in venules or large veins (Stewart et al. 1974).

The platelets, especially when in the "adhesive state" with long pseudopods are unique in sticking to the wall provided it is itself adhesive. The adhesive state is difficult to define: intact endothelium is not in such "adhesive state". Subendothelial structures as well as rough glass and polymer surfaces and on a molecular basis fibrinogen adsorption appears to be pivotal (Zucker and Vroman 1969).

We can deduce the following conditions for wall adhesion from our experiments. Prerequisite for this process is a high platelet density near the wall and a frequent encounter - which both are seen only in the presence of red cells. Triggering events are derived from wall properties such as any kind damage (e.g. endothelial detachment) especially when associated by filtration (as discussed by Forstrom et al. 1975), wall protrusion and deposited platelets - especially when associated with the formation of platelet pseudopodia that reach into the fluid. It follows from simple fluid-dynamic considerations that any flow vector directed towards the wall would augment the above mentioned events proportional to the convective flux of blood. Lastly, there are enhancing mechanisms which include various biological properties of platelets, e.g. the release of chemical mediators and the demasking of platelet phospholipids. Mechanical effects by deposited platelets include possible entrapment of passing platelets or a filtering effect of a mechanically loose platelet aggregate that allows passage of plasma but traps platelets coming from upstream. Downstreams from an existing platelet deposit, secondary flow and reattachment point may further augment platelet deposition and aggregation. It goes without saying that all of these processes can be further accelerated in case the platelets circulate in "activated" or even aggregated form or when they are locally "activated" by fluid dynamic forces (see Wurzinger et al., this volume). Finally, one can further predict that the processes described above are limited by the growth of the deposit, by smoothing processes or by an effect on the pressure drop that leads to shear stresses high enough to disrupt deposited platelets.

The results of our work can be summarized as follows:

1. Irrespective of the wall properties, platelets stick to it only provided that they are previously "activated" and are brought to it by a convective flow such appears in stagnation point flow, re-attachment points or when platelets hit the wall protrusions.
2. The presence of red cells drastically increases the concentration of platelets in the peripheral layer immediately adjacent to the wall: Consequently, in rapid flow of a platelet-red cell mixture the transport rate of platelets within that critical layer can be more than 70times higher than in flow of platelet rich plasma.

We consider these microrheological findings important for all forms of deposition thrombosis and the conclusion was derived from experiments which in many points corroborate the findings of the pioneering work of Goldsmith and Karino.

References

Antonini G, Guiffant G, Quemada D, Dosne AM (1978) Estimation of platelet diffusivity in flowing blood. Biorheology 15: 111-117

Baldauf W, Wurzinger LJ, Kinder J (1978) The role of stagnation point flow in the formation of platelet thrombi on glass surfaces in tubes with various geometry. Path Res Pract 163: 9-33

Begent N, Born GVR (1970) Growth rate in vivo of platelet thrombi, produced by iontophoresis of ADP, as a function of mean blood flow velocity. Nature 227: 926-930

Bennet L (1968) Blood flow: velocity profile continuity near a wall. Biorheology 5: 253-262

Bergqvist D, Arfors KE (1977) The role of red cells in haemostatic plug formation in the isolated rabbit mesentery. Thromb Res 11: 95-100

Feuerstein IA, Brophy JM, Brash JL (1975) Platelet transport and adhesion to reconstituted collagen and artificial surfaces. Trans Am Soc Artif Intern Organs 21: 427-434

Fischer TM, Schmid-Schönbein H (1977) Tank tread motion of red cell membranes in viscometric flow: behaviour of intracellular and extracellular markers (with film). Blood Cells 3: 351-365

Fischer TM, Stöhr-Liesen M, Schmid-Schönbein H (1978) Red cells as a fluid droplet; tank tread like motion of human erythrocyte membrane in shear flow. Science 202: 894-896

Forstrom RJ, Bartelt K, Blackshear PL, Wood T (1975) Formed element deposition onto filtering walls. Trans Am Soc Artif Intern Organs 21: 602-607

Goldsmith HL, Yu SSK, Marlow J (1975) Fluid mechanical stress and the platelet. Thromb Diathes Haemorrh 34: 32-41

Keller KH (1981) The dynamics of the interaction of cells with surfaces. In: Salzman EW (ed) Interaction of the blood with natural and artificial surfaces. Marcel Dekker, New York Basel, p 119

Kinder J, Kratzer M (1975) Geschwindigkeitsmessung im Inneren komplizierter Strömungen mit einem Lichtschnittverfahren. Biomed Technik 20: 11-12

Leonard EF (1972) The role of flow in thrombogenesis. Bull NY Acad Med 48: 273-280

Marzec U, Johnston GG, Bernstein EF (1975) Platelet function in calves: a study of adhesion, aggregation, release, clotting factor activity and life span. Trans Am Soc Artif Intern Organs 21: 581-585

Müller-Mohnssen H (1971) Pathogenese der Koronarsklerose und Strömungsmechanik. Münchner Med Wschr 113: 604-616

Richardson PD (1973) Effect of blood flow velocity on growth rate of platelet thrombi. Nature 245: 103-104

Schmid-Schönbein GW, Skalak R, Usami S, Chien S (1980) Cell distribution in capillary networks. Microvasc Res 19: 18-44
Schmid-Schönbein H, Born GVR, Richardson PD, Cusack N, Rieger H, Forst R, Rohling-Winkel I, Blasberg P, Wehmeyer A (1981) Rheology of thrombotic processes in flow: the interaction of erythrocytes and thrombocytes subjected to high flow forces. Biorheology 18: 415-444
Stewart GJ, Ritchie GM, Lynch PR (1974) Venous endothelial damage produced by massive sticking and emigration of leukocytes. Am J Path 74: 507
Turitto VT, Baumgartner HR (1974) Effects of physical factors on platelet adherence to subendothelium. Thromb Diathes Haemorrh [Suppl] 60: 17
Turitto VT, Baumgartner HR (1979) Platelet interaction with subendothelium in flowing blood: effect of blood shear rate. Microvasc Res 17: 38-54
Wolf MK, Aronson SB (1961) Growth, fluorescence and metachromasy of cells cultured in the presence of acridin orange. J Histochem Cytochem 9: 22-29
Wurzinger LJ (1979) Hydrodynamisch induzierte Plättchenablagerungen an Glasmodellen verzweigter, gekrümmter und verengter Gefäßabschnitte und Speziesunterschiede im Plättchenaggregationsverhalten von Mensch, Rind, Schwein, Schaf, Hund, Kaninchen und Truthahn. Dissertation RWTH Aachen
Zucker MB, Vroman L (1969) Platelet adhesion induced by fibrinogen adsorbed onto glass. Proc Soc Exp Biol Med 131: 318

Discussion

FRIEDMAN: Have you done any of the platelet deposition experiments in unsteady flow, and if not, do you intend to? Second, could you elaborate on your statement in which you mentioned the non-uniformity of the hematocrit in the vessel and how that might influence platelet deposition?

WURZINGER: To answer Dr. Friedman's first question about the platelet deposition experiments in branchings and bends, we did only a few experiments with pulsatile flow. We found basically the same deposition pattern as in steady flow, although the quantity of deposited platelets was somewhat lower.

BLASBERG: To comment on the non-uniformity of hematocrit, this effect depends on the cell species at which you are looking. Whereas the concentration of red cells is higher in the axial stream than in the marginal layer, platelets occur at a higher concentration in the wall adjacent layer than in the axial core. Furthermore, the presence of red cells augments the motion of platelets toward the vessel wall, thus increasing the possibility of platelet deposition. One explanation may be that the average spacing of the red cells in the central stream is so small that platelets are excluded to a certain degree.

BLACKSHEAR: If it is true that the platelets enter the tube as well mixed blood, then there should have been an increase in platelet concentration near the wall in the developing flow in the tube. Did you have occasion to look at the platelet concentration along that length?

BLASBERG: The suspensions used were mixed by a magnetic stirrer before entering the tube, but we looked at the platelet distribution only after an entrance length of one hundred times the tube diameter at minimum.

BLACKSHEAR: But that brings up a question as to whether there is any tube in which platelets do not feel crowded and therefore should be in the wall everywhere no matter what size tube you have?

BLASBERG: We think that there may be certain flow situations in the circulation in which the blood will be mixed very well, and the reported concentration differences between red cells and platelets will vanish until the non-uniform distribution is achieved again.

DEWEY: I do not know what the extent of the displacement of the red cells is at the wall, but it occurs to me that if you get too large a layer, the viscosity is going to change.

BLASBERG: There are a few red cells close to the wall in the range of shear rates we have used; no doubt, there is a skimming layer containing only a few red cells and many platelets.

The Localization of Transient Subendothelial Water-Filled Blisters in the in Situ Rabbit Aorta as a Result of Reduction in Pressure

P. L. Blackshear, Jr., G. L. Blackshear, M. K. Newell, S. J. Kayser, and P. F. Emerson

Introduction

Our group at the University of Minnesota has been measuring the water flux across the vascular intima in dog and rabbit aortas with steady and time varying pressures. We have found evidence of reverse fluid movement from wall to lumen when arterial pressure is lowered suddenly.

This evidence reported earlier in the proceedings of the preceding meeting of this present symposium [1] included measurements of pressure by needles placed in the media. These showed that a driving force is produced by a pressure change in the lumen. In addition, direct measurements of transient reverse flow from media to lumen, in an air inflated arterial segment in which a falling fluid film bathed the intima, supported the hypothesis that a transient stress generated pressure gradient in a fluid filled porous media produced a flow that obeyed Darcy's law locally and instantaneously. On the basis of these findings, we anticipate that reverse flow that accompanies a postural change could cause a substantial fluid displacement.

We expected the effect of reverse fluid flow in the arterial wall would involve loosening of endothelial cell bonding to the substratum, and were interested in the forces that held these cells to the walls in health and deviations from health.

In particular, we wished to know if it were possible to fix the artery following a maneuver designed to loosen these bonds and infer anything quantitative about fluid movement that would help in predicting the role of transmural water movement in plaque formation.

The anatomy of the intimal layer of arteries is quite variable in different segments of the vascular tree as well as in different species. The sites in the aorta which are prone to atherosclerosis in man tend to have thin basement membranes, an intimal cushion that increases in thickness with age, and a relatively avascular media with thick laminae. The endothelial cells form a tight mosaic lining and the cells have complicated intercellular junctions. It has been suggested that loss of cells from this lining layer occurs with various mechanical and chemical stresses [1-9] and that this loss tends to be focal and to occur at points where hemodynamic disturbances exist.

A counter observation made at this symposium is that denudation of the subendothelium is a rare event and that the plaques occur in the intima and media beneath a continuous endothelial layer. An important question is why these cells which form the interface between blood and arterial wall will at times either slough off and leave an area of wall that must be repaired or invite disturbances by other means. Others have shown that minimal injury frequently involves subendothelial edema and that the endothelial cells are more or less loosened from their substrate and "hang on" by means of occasional tethers, forming essential-

Fluid Dynamics as a Localizing Factor for Atherosclerosis
Editors: G. Schettler et al.

ly a blister on the arterial lumenal surface. In fact, such blister-like lesions have been reported by pathologists as an early lesion of atherosclerosis.

The results of our and other recent studies gives the following generalizations for steady flow. The resistance for normal water movement across the intima is partly endothelial and partly medial resistance. The medial resistance is thought to be that of a porous medium of water permeability of $k'/\mu = 10^{-9}$ $cm^2/s \cdot cm\ H_2O$ (where k' is the Darcy constant and μ the fluid viscosity), and a half thickness at physiological pressures of $\delta = 2 \cdot 10^{+2}$ μm and less [10]. This represents the distance from the lumen to the nearest layers of vasa vasorum. The total media thickness, $2\delta'$, is roughly proportional to the artery radius, r [11]. When δ' becomes less than 200 μm, the thickness of the unvascularized layer of the media, 2δ, δ becomes equal to the total media half thickness and thus is proportional to vessel radius for the smaller vessels.

In experiments performed in rabbit, dog, and pig aortas, from which the endothelium has been removed, the water flux obeys Darcy's law with a permeability of $k/\mu = 10^{-9}$ $cm^2/s \cdot cm\ H_2O$ [12,13,14]. Thus the hydraulic conductivity, k/μ, of the medial tissue of an artery increases as the radius decreases. Measurements with and without endothelium on rabbit aortas gives the value for the hydraulic conductivity for the endothelial layer alone to be $L_P = 8.6 \cdot 10^{-8}$ $cm/s \cdot cm\ H_2O$. Because this value is so close to the $L_{PE} = 8.6 \cdot 10^{-8}$ $cm/s \cdot cm\ H_2O$ reported for the rabbit heart capillary and $L_{PE} = 10 \cdot 10^{-8}$ for muscle capillaries [14] it seems reasonable to suppose that the arterial tree from heart to muscle capillaries has a $L_{PE} = 9 \cdot 10^{-8}$ $cm/s \cdot cm\ H_2O$. Then the overall hydraulic conductivity L_P found from $1/L_P = 1/L_{PE} + 1/L_{POL}$ is half the capillary value for arteries of 3 mm diameter or greater; Lp tends to the capillary value as artery size diminished and, depending on position of the vasa vasorum, tends to the outer layer value as the size increases.

These values have been obtained in vitro using a perfusion medium containing 1% protein. They suggest a more permeable endothelium than inferred from in vivo dye penetration experiments [15,16], Recently, a study on albumin concentration effects on capillary L_P [17] suggests that concentration polarization of macromolecules could account for this discrepancy.

The driving force for transmural water movement fluctuates during the day and is influenced by body attitude. On average there is a gross flow out of the arterial-capillary tree of 20 l/day and a net flow from the vascular tree to the lymph of 2-4 l/day. Upon rising from prolonged recumbency there is superimposed on this average movement a further volume shift from the upper body tissue into the vascular tree, and a still greater volume shift from the vasculature into the tissue space in the lower extremities and vice versa on lying down. Swings in blood osmolarity can also cause transmural shifts in fluid volume. Considerations of the fluid filled tissue space within the arterial media led Kenyon [17] to predict that this space would inflate when arterial pressure increased and deflate when it fell.

In the Physiology-Bioengineering Laboratory at the University of Minnesota, we have been measuring transmural water movement directly in vitro, in response to a pressure transient and have sought evidence of water movement due to imposed pressure transients in vivo. In this report, we summarize the in vitro measurements of transient reverse flow in dog aortas and the results of recent attempts to induce blister formation in rabbit aortas in situ following a pressure fall, as a means of quantitating the fluid displacement.

Methods

Transient Velocity Measurements. The dog descending thoracic aorta is mounted at its normal length in a fluid filled cup. Cannulae permit air inflation and a fluid spray forming a film of ca. 0.025 cm thick on the inner wall, Fig. 1. Fluid is carried by the exit cannulato a restricted drain which maintains the pressure high with a minimum flow. The steady state weight increase yields the mean hydraulic conductivity, L_p $v/\Delta p$, where v is mean velocity and Δp the pressure drop.

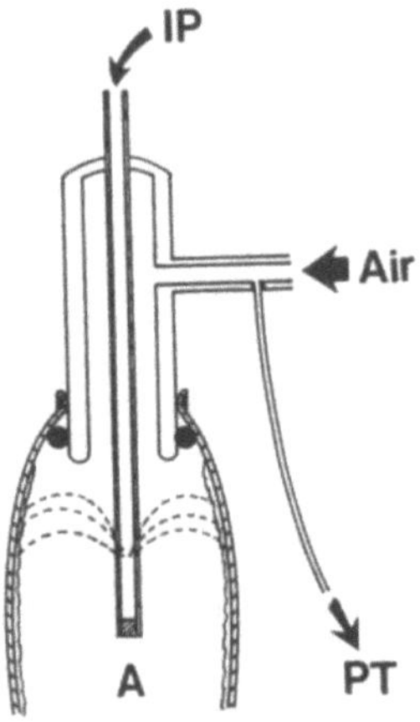

Fig. 1. Detail of internal perfusate, IP, introduction into the artery, A, and the location of the lead to the pressure transducer, PT. The rate of weight increase of the entire specimen following a pressure transient is used to infer the v´ at the lumenal surface. The artery is enclosed in a fluid sack so that fluid movement across the outer wall does not influence the weight

The transient in pressure produces a transient weight gain which is used to determine a transient hydraulic conductivity, $L'_p = v'/p'$, where v' is perturbation velocity and p', the perturbation pressure. The values of v' reported below were obtained after a series of pressurizations and relaxations were used to condition the artery so that results could be repeated.

In Situ Blister Formation. Preliminary studies in a segment of the thoracic aorta of rabbits cannulated in situ have been conducted in which part of the arterial segment was exposed to a sudden drop in pressure (to 25 cm H_2O from its precannulation value) and part kept at constant pressure and subjected to 10 minutes perfusion with oxygenated Krebs Ringer solution followed by 2 minutes of Evans blue dye, followed by a glucose wash and silver nitrate staining under these conditions. The perfusion was controlled over a range of shear rates. In one series of experiments a steady perfusion of an arterial segment was performed employing a distal cannula designed to expose the endothelial layer to a locally high fluid shear.

Results and Discussion

In Vitro Studies. Transient data (primed symbols) are summarized in Fig. 2. Note that when $p' < o$, $v' < o$ and therefore $L'_p = v'/p' > o$.

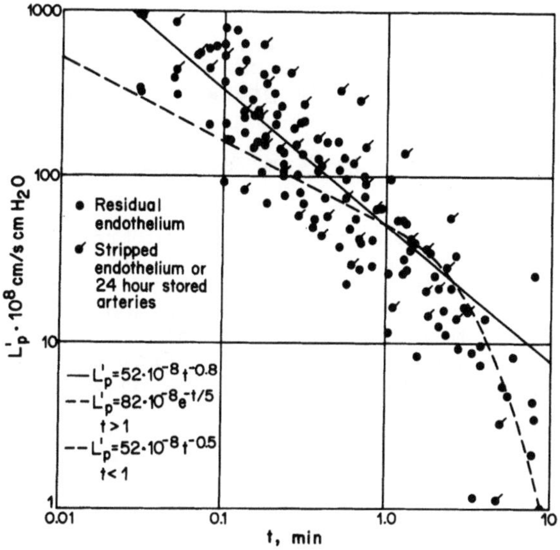

Fig. 2. Log-log plot of all the L'_p's

If we apply Kenyon's simple consolidation theory [18] to the data, we obtain a best fit for the t 1 minute data with the dashed line in Fig. 2, shown in dimensionless form in Fig. 3. Here p_t is the inferred tissue pressure perturbation and D is the diffusion coefficient governing tissue pressure transients according to the diffusion equation [13, 18], and is related to the specific compliance of the unvascularized part of the media, $C_S = k/\mu\ D$.

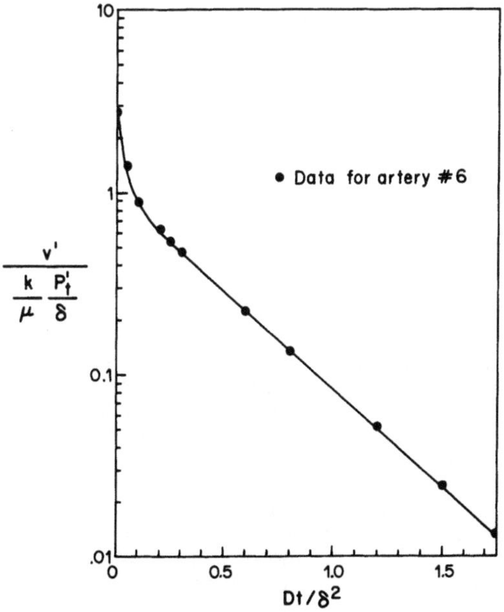

Fig. 3. Solutions from the conduction equation [15]. Semilog plot (data from a single artery included)

C_s, in turn is related to D_p, the integral of v' from $t = o$ to $t = \infty$, divided by the half thickness of the avascular media, δ, and the tissue pressure changes p_t; $C_s = D_p/\delta p'_t$. A similar value can be obtained by measuring a weight loss when an arterial media is pressed against a support screen [12].

In [13] we obtain $p_t = -5.15\ p'$ and $D = 2 \cdot 10^{-6}\ cm^2/s$. Thus $C_s = 8 \cdot 10^{-4}\ (cm\ H_2O)^{-1}$.

At present we feel that both the tissue compliance, C_s, and transient tissue pressure to perturbation pressure ratio, p_t/p', will increase with increased hydration and will be a function of the state of stress of the artery. An example of stress dependence can be obtained by comparing the value to the Harrison and Massaro data [12] which gives $C_s = 2.10 \cdot 10^{-3}$ when a slice of pig aorta is compressed from a relaxed state, recall our inflated artery yields $C_s = 8 \cdot 10^{-4}\ (cm\ H_2O)^{-1}$.

Values extracted from the literature for other tissues are compared in Table 1. The quantity of fluid available to lift the endothelial layer is $D_{p\ net}$, the integral of v' from $t = o$ to $t = t_R$, the time when the velocity at the wall reverses. $D_{p\ net}$ divided by D_p, from the simple theory, is of order unity. Measured $D_{p\ net}$ values are comparable [13] and are of order 10 µm for the p_t imposed. To put these values of D_p into perspective, a fully inflated SEV has a $D_{p\ net} \leq 2$ µm without blister formation [17].

Table 1

Tissue	Specific Compliance $(cm\ H_2O)^{-1}$		Source
Rabbit myocardium	0.06	Blackshear	19
Rabbit myocardium	0.10	Vargas and Blackshear	20
Dog leg muscle			
low p	$8 \cdot 10^{-2}$	Guyton	21
high p	$8 \cdot 10^{-3}$	Guyton	21
Pig aorta (relaxed)	$2 \cdot 10^{-3}$	Harrison and Massaro	12
Dog aorta (physical extension)	$8 \cdot 10^{-4}$	Blackshear, Vargas, and Blackshear	13
Articular cartilage	$5 \cdot 10^{-5}$	Malcom	22

In Situ Studies. Results of the preliminary studies involving Evans blue dye and the high shear stress cannula showed the following behavior: blisters formed; they were torn off by the local (high) shear stress in some instances and in others the stress simply caused endothelial injury as evidenced by clusters of cells with dyed nuclei. The lack of general subendothelial dye uptake lead us to conclude that at the time of introduction of the dye, the endothelial layer was not excessively permeable. This conclusion is based on the absence of blue staining in segments maintained at a shear stress of approximately 10 dynes/cm^2 or less. Blisters are always generated in the high velocity region of the distal cannula inlet which is especially designed to produce a local high-velocity shear stress ($\simeq$ 100 d/cm^2) region at the wall. When the fixation occurs at 100 cm H_2O there are streaks of blisters of endothelial cell size or smaller in the high shear regions. When fixation is 10 minutes after exposure to a sudden drop from 100 to 25 cm H_2O pressure (and lower shear stress), the endothelial layer is locally removed over the stressed region.

At this point a tentative qualitative relationship between the local fluid mechanics and the local artery wall mechanics suggests itself:

At fluid shear stress greater than 500-1000 d/cm^2 the undisturbed endothelium can be removed [23]. For fluid stress less than 500 d/cm^2, the endothelium remains attached but shows an enhanced permeability [24,25]. If a change in hydrostatic pressure has caused an upwelling of fluid from beneath the endothelium such that blisters form and grow in size, there appears to be a critical size for a given shear rate at which the blister will tear off. This size becomes smaller as the shear stress approaches 500 d/cm^2.

Such small denuded patches would be expected to be filled in by near neighbors without serious consequences. If the fluid shear stress is sufficiently low, the blisters won't tear before they subside. These qualitative considerations suggest that siting of plaques by blister rupture would occur at an intermediate shear stress that is small enough to allow large blisters to form, so that the blister scar would be large enough to require cell mitosis to cover. The stress must, however, be large enough to cause the blisters formed to be ruptured before they can subside.

Blistering Dynamics. The literature on blistering of elastomeric films is generally focused on the thermodynamics or fracture mechanics but not on the dynamics [26,27,28].

In the following derivations we explore dynamic aspects of blister formation in an effort to derive information on fluid movement from blister size and spacing. We model the endothelial layer as a sheet with negligible bending resistance at the blister size level and infinite resistance to bending at the tether spacing level. Elastomeric tethers obeying Hook's law anchor this sheet to a rigid but porous subendothelium through which water can move. This model is based on the micrographs of Tsao and Glagov [29].

When displacement, y, exceeds a critical displacement, δ_c, owing to fluid pressurization of subendothelial volume, SEV, tethers break and fluid moves from the surrounding SEV to inflate a blister. In the model, the blisters are circular when viewed en face.

The subendothelial volume SEV, is considered completely deflated prior to the imposition of a transient in fluid movement. A necessary condition for subendothelial volume swelling is that there be a net flow into the SEV. Thus, the transient flow per unit area, v', entering the SEV from the media, induced by a sudden drop in luminal pressure, $-p'$, would have to equal the losses from the SEV which we take to be composed of the steady component, i.e. the steady filtration rate across the media, plus the osmotically driven drainage through the endothelial layer. Because $v' \to \infty$ at $t = o$, this condition is always met: There will be some SEV swelling following any sudden reduction in luminal pressure (see Table 2).

Similarly, in addressing the necessary conditions for blister formation, we determine the conditions required to achieve an inflation of the SEV to a displacement of $\delta_c = 1$ µm. This requires a $|p'| \geq 0.25\ \Delta P$, i.e. a $p_t \geq \Delta P$.

Table 2. Necessary condition for subendothelial volume swelling

SEV swells when $v_{net} > o$

$$o < v_{net} = 0.3\ k\ \ p_t/(\mu\delta)(Dt/\delta^2)^{-1/2} \underbrace{- k\ \Delta P/(\mu 2\delta)} - L_{PE}\ \underbrace{(P_{SEV} - P_L - \pi_{SEV} + \pi_L)}$$

For rabbit thoracic aorta	$(4 \cdot 10^{-6})$	$(2 \cdot 10^{-6})$	cm/s

for SEV to swell $0.3\ k\ p_t/(\mu\delta)(Dt/\delta^2)^{-1/2} < 6.10^{-6}$ cm/s

Next we seek to learn what can be inferred from the spacing of blisters in a blister field. Such a field is shown in [30], see Fig. 4.

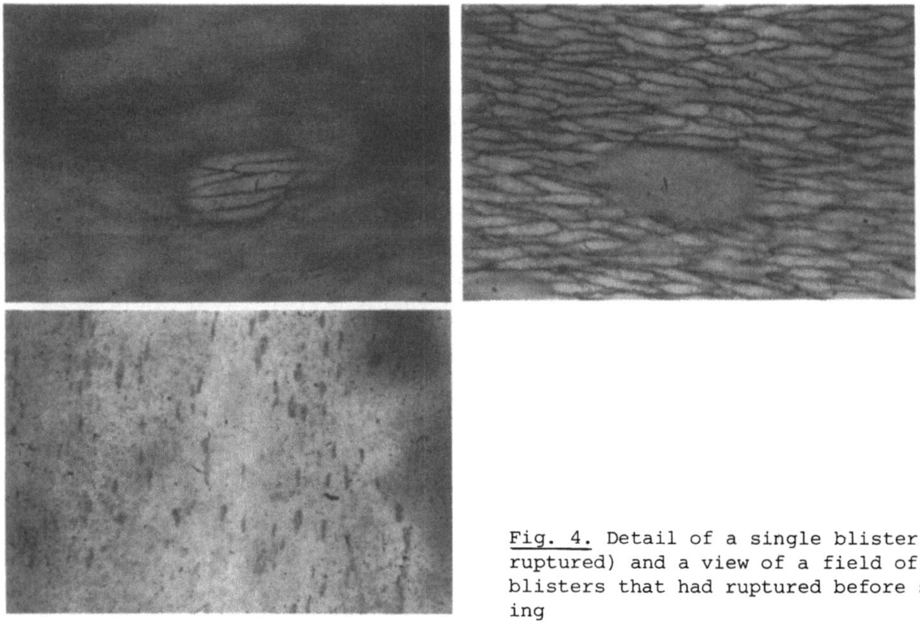

Fig. 4. Detail of a single blister (unruptured) and a view of a field of blisters that had ruptured before staining

We model this field as one in which circular blisters of radius R_1 form in the center of hexagons of uniform size. We approximate the hexagons with a circle of radius R_2 (see Fig. 5).

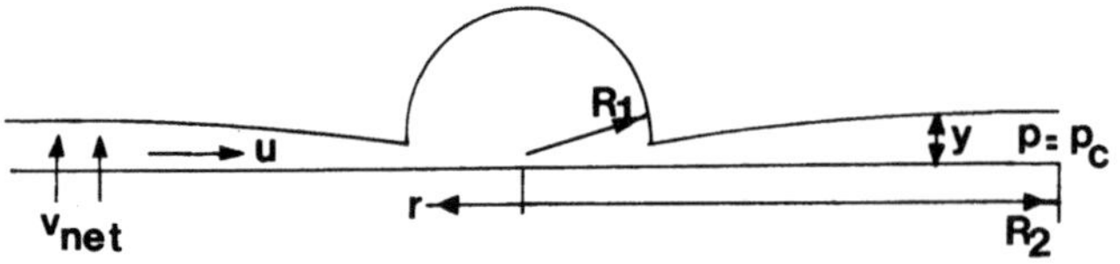

Fig. 5. Model of the blistering process; a single blister of radius R_1 will stress-relieve the SEV for a region of diameter R_2 where the net fluid in-flow is v_{net}

After the blister has been produced, we assume the pressure inside the blister is the lumenal pressure which we take to be zero. At R_2, we assume the pressure in the SEV $p = p_c$, the pressure required to displace the endothelial layer a distance δ_c, the maximum displacement allowed before tethers break and blisters form. There is a net inflow at each radius, r, determined by the difference between the flux into the SEV from the media and the flux out of the endothelial layer. A sketch of the cross section of this SEV is shown in Fig. 5. Because the pressure at R_2 is greater than that at R_1, there will be a flow with mean velocity at r; we may relate u to the pressure gradient by Darcy's law

$$u = (k'/\mu)dp/dr \quad , \tag{1}$$

where

$$k' = y^2/12 \tag{2}$$

the value for Poiseuille flow in a slit of total width y. y in turn is assumed to be related to the local pressure by

$$y = \delta_c \, p/p_c \quad . \tag{3}$$

Then from continuity if v_{net} = const

$$2\, ryu = (R^2 - r^2)v_{net} = 2\, y(k'/\mu)dp/dr \quad . \tag{4}$$

Then integrating (4) and having expressed y and k' as functions of p, we obtain

$$v_{net} = \delta_c{}^3 p_c/(24\mu R_2^2\, I) \quad , \tag{5}$$

where

$$I = \ln\,(R_2/R_1) - (1 - R_1^2/R_1^2)/2 \quad . \tag{6}$$

The bounds of I range from a high of 6.4 for a single 10 μm blister in an artery of 1 cm circumference to 1 for closely spaced blisters.

Thus the model suggests that the dimensionless quantity:

$$p_c(\delta_c/R_2)^2/\mu v_{net}/\delta_c = 24\ I,\ 6 \leq 6\ I \leq 152 \tag{7}$$

i.e., 6 I is order of magnitude of 40.

We will refer to the reciprocal of this quantity as the blister number, B.N.

Our interpretation of this analysis is that when the blister number $\mu v_{net}/\delta_c/p_c(\delta_c/R_2)^2$ is less than, say 0.006 a single blister would stress relieve the entire artery while for a blister number greater than 0.006 (say B.N. = 0.025) blisters of spacing R_2 < artery circumference will be observed.

We don't know the value of p_c but feel it should be quite close to 500 d/cm^2, the approximate critical shear stress for endothelial cell removal found by Fry [23], δ_c can be estimated from the micrographs of Tsao and Glagov [29], $\delta_c = 10^{-4}$ cm; in addition $\mu = 0.01$; then with 6 I = 40, i.e. the blister number = 0.025 leads to

$$v_{net} = 1.25 \cdot 10^{-9}/R_2^2 \quad .$$

For the larger (approximately 3.5 kg) rabbits we always get blistering with $R_2 \approx 10^{-2}$ cm when we suddenly lower the pressure ($p' = -40$ cm H_2O) whereas for the 2 kg rabbits we never observe blistering. Because, see Table II, the other contributors to v_{net} (besides v'), are small compared to $1.25 \cdot 10^{-5}$ cm/s; we may take $v_{net} = 1.25 \cdot 10^{-5} = v_b'$, the value of v' at the instant the blisters form.

Such a high velocity at blistering must occur at a time $t_b < 0.2 \times (\delta^2/D)$ when D is the diffusion coefficient governing consolidation in the porous medium. Under these circumstances, we may approximate the dependence of v' and y vs time for a porous slab in which the interstitial pressure is suddenly elevated an amount p_t at t = o. Then for $t < 0.2\ \delta^2/D$

$$v' = 0.3\ k\ \ p_t/(\mu\delta)(Dt/\delta^2)^{-1/2} \tag{8}$$

$$y = [k\ p_t\ \delta^2/(\mu\delta D)]\ (Dt/\delta^2)^{1/2} \tag{9}$$

Setting $v' = 1.25 \cdot 10^{-5}$, $y = 10^{-4}$:

$t = t_b = 2.4$ s

Since $D = 6 \cdot 10^{-6}$ cm^2/s

and $k\mu/D = 8 \cdot 10^{-4}$ (cm H_2O)$^{-1}$

then $p_t = 33$ cm H_2O; but $-p' = 40$ cm H_2O

$\therefore\ p_t/-p' = 0.8 = a$

According to consolidation theory a should be radius of artery/$\delta \approx 5$. The value of a determined from dog arteries modeled by consolidation theory earlier is 5.15 [13] while a pure conduction model requires $a \simeq 1$.

This agreement encourages us to believe that where blisters appear, their size and spacing can be used to infer the value of v' at the instant of blistering and from this with the help of consolidation theory we may determine the parameters that characterize transient fluid movement in response to a pressure change.

There are many regions in the 3.5 kg rabbits where no blisters are observed. In rabbits smaller than 2 kg, we never observe blisters. We infer from the model that when no blistering occurs, the blister number leading to Eq. (7),

$$(\mu v'/\delta_c)/P_c(\delta_c/R_2)^2 \text{ is less than } 0.006$$

where R_2 is taken to be half the circumference of the vessel.

The absence of blistering in the small rabbits, (if our assumed values for p_c and δ_c holds) means that the blister number is less than 0.006 or $v_{net}' = 10^{-9}$, a velocity much smaller than even the steady transmural velocity. See Table 2. In addition, y is throughout less than δ_c.

Why, then does y not exceed δ_c when p' is large? We infer that it is because p_t is too small, i.e. in the equation

$$p_t = -a\ p'$$

a is too small. Anisotropy in the arterial wall could account for this as could an interstitial fluid displaying viscoelastic properties.

In the limit that Poissons ratio for the arterial tissue is such that p_t is influenced only by conduction from the surfaces, $p_t \leqq -p$ and $a \leqq 1$, the value of a should be smaller the lower is L_{PE} and the smaller δ. This interpretation would explain why the smaller (younger) rabbits did not show blisters while the larger (older) ones did. The inferred value of compliance C_s in Table 1 would be higher and the Eqs. (8) and (9) modified somewhat. Such a "pure conduction" model should be explored further.

Finally, what does all this have to do with fluid mechanics and atherosclerotic plaque sitings?

As we have seen, it is possible to produce blisters in the endothelial layer by reducing the pressure suddenly. These blisters would presumably subside with no injury to the subendothelium if the local shear stress is sufficiently low. In [30], we show ca. 100 µm long blisters that underwent desquamation at 4 dynes/cm^2. Smaller blisters presumably would not be torn away at this shear stress and these blisters would not have broken if the local shear stress were lower. The blister would not have grown that large if the shear stress were higher. Indeed, Nerem [24] and Thibault [25] have shown that at high shear stress endothelial permeability increases and thus v'_{net} would be reduced. Small blisters when broken should produce lesions that can easily be covered by the migration of neighboring cells. These considerations suggest that if the arterial tissue and pressure variations provide the conditions for blister formation, the size of the lesion caused would, if rupture occurs, vary with shear stress. The general regions of the vasculature where blisters might be expected are where large pressure swings and thick arterial walls may be found. We would expect a regional distribution as shown in Fig. 3 of [30]. An example of the local blisters occurring at intermediate shear stress regions are frequently seen at the ostium of a rabbit intercostal. Glagov has earlier [31] published evidence of lesions in this region in cholesterol fed rabbits.

Thus, there appears to be a plausible model for a fluid mechanical phenomenon that influences blister formation and subsequent plaque formation at particular sites. These sites must undergo hydrostatic or osmotic pressure swings (of ca. 40 cm H_2O) for duration of seconds and they must have a media fluid content that produces a perturbation velocity sufficient to cause blisters. Finally, local fluid shear stress should be intermediate; too high a stress will open blisters when they are small and the repair process would be rapid; to low a shear stress will not rupture the blisters.

References

1. Payling-Wright H (1973) Endothelial injury and repair. Bibl Anat 12: 87-91
2. Tasao C (1970) Graded endothelial injury of the rabbit aorta. Arch Path 90: 222-229
3. Weber G, Losi M, Toti P, Vati R (1979) Circulation endothelial-like cells in arterial peripheral blood hypercholesteremic rabbits. Artery 5: 29-36
4. Gerrity RG, Richardson M, Aplan BA, Cade JF, Hirsch J, Schwartz CJ (1976) Endotoxin-induced vascular endothelial injury and repair. II. Focal injury, en face morphology, ^{3}H thymidine uptake and circulating endothelial cells in the dog. Exp Mol Path 24: 59-69
5. Payling-Wright H (1973) Endothelium: Recovery of cells from circulating blood and subsequent repair. In: Cavallero C (ed) Artery wall in atherogenesis. Piccin Medical Books, Padua, pp 13-23

6. Shimamoto T (1963) The relationship of edematous reaction in arteries to atherosclerosis and thrombosis. J Atheroscler Res 3: 87-1102
7. Trillo AA, Prichard RW (1979) Early endothelial changes in experimental primate atherosclerosis. Lab Invest 41: 294-302
8. Constantinides P, Robinson M (1969) Ultrastructural injury of arterial endothelium. II. Effects of vasoactive amines. Arch Path 88: 106-117
9. Gutstein WH, Furrell GA, Armellini C (1973) Blood flow disturbance and endothelial cell injury in preatherosclerotic swine. Lab Invest 29: 134-149
10. Vargas CB, Vargas FF, Pribyl JG, Blackshear PL Jr (1979) Hydraulic conductivity of the endothelial and outer layers of the rabbit aorta. Am J Physiol 236: H53-H60
11. Wolinski H, Glagov S (1967) Nature of species differences in the medial distribution of aortic vasa vasorum. Circ Res 20: 409-421
12. Harrison RG, Massaro TA (1967) Water flux through porcine aortic tissue due to a hydrostatic pressure gradient. Atherosclerosis 25: 363-367; and Harrison RG (1976) Ph.D. Thesis, University of Wisconsin
13. Blackshear PL Jr, Vargas CB, Blackshear GL (1981) Artery wall movement related to atherogenesis. Submitted for publication
14. Samje LH, Zweifach BW, Intaglietta M (1970) Micropressures and capillary filtration coefficients in single vessels of the cremaster muscle of the rat. Microvasc Res 2: 96
15. Colton CK, et al. (1980) Labeled albumin transport into the normal and deendothelialized rabbit thoracic aorta in vivo. In: Nerem R (ed) Hemodynamics and the artery wall. Proceedings of a Specialists Meeting Sponsored by NSF, Houston, TX
16. Blackshear GL, Vargas FF, Blackshear PL Hydraulic conductivity of rabbit heart capillaries determined by two independent methods. Paper in preparation
17. Kenyon DE (1979) A mathematical model for water flux through aortic tissue. Bull Math Biol 41: 79-90
18. Vargas FF, Blackshear GL (1981) Transcapillary osmotic flows in the in vitro perfused heart. Am J Physiol 240 (Heart Circ Physiol 9): H448-H456
19. Blackshear GL (1975) The effect of hypoxia and histamine on the capillary conductance of the rabbit heart. Ph.D. Thesis, University of Minnesota
20. Blackshear GL, Vargas FF Hydraulic conductivity of rabbit heart capillaries determined by osmotic and hydraulic methods (in preparation)
21. Guyton AC (1972) I. Compliance of the interstitial space and the measurement of tissue pressure. Pflügers Arch 336: S1-S20
22. Malcolm LL (1976) Frictional and deformational responses of articular cartilage interfaces and dynamic loading. Ph.D. Thesis, UCSD, La Jolla, CA
23. Fry DL (1976) Certain histological and chemical responses of the vascular interface to acutely induced mechanical stress in the aorta of the dog. Circ Res 24: 363-367
24. Nerem RM, Cornwall JF (1980) The role of fluid mechanics in atherogenesis. J Biomech Eng 102: 181
25. Thibault LE, Fry DL (1980) Hydrodynamically induced wall shear stress effects on Evans blue dye uptake. In: Nerem RM (ed) Hydrodynamics and the artery wall. Proceedings of Specialists Meeting Sponsored by NSF, Houston, TX
26. Dannenberg H (1961) Measurement of adhesion by a blister method. J Appl Polym Sci V: 125-134
27. Meer-Lerk LA van der, Heertjes PM (1975) Blistering of varnish films on substrates induced by salts. J Oil Col Chem Assoc 58: 79-84
28. Meer-Lerk LA van der, Heertjes PM (1981) The influence of pressure on blister growth. J Oil Chem Assoc 64: 30-38
29. Tsao C, Glagov S (1970) Basal endothelial attachment tenacity at cytoplasmic dense zones in the rabbit aorta. Lab Invest 23: 510
30. Blackshear PL Jr, Blackshear GL, Newell MK, Emerson PF, Kayser SJ (1982) Gross endothelial layer blistering and vascular injury. To be published in Proceedings of 6th Int Symp on Atheroscler, June 13-17, Berlin, FRG
31. Glagov S, Clark JM A unitary view of arterial medial microarchitecture. Ibid, p 52

Discussion

DEWEY: There is something that I am missing with regard to the "lying down" experiment. You are saying that, if the artery has somehow become accomodated to the ordinary hydrostatic pressure, then when you lie down there would be enough pressure to cause blistering? What happens in exercise? I mean if somebody jumps or something. You have pressures generated by the acceleration of the blood in the arterial column that can cause many times that force.

BLACKSHEAR: First, the characteristic time is 200 seconds for the blistering to occur, so that something like a high pulse or sudden jolt would not bother you. It would have to be a prolonged insult. I know what would happen if human arteries behave as these excised dog arteries behave, and from our rabbit data we infer numbers that are not too different from the dog data. However, the human intima is a lot thicker than that in the rabbit, so it may be a lot more forgiving.

CARO: I have problems. Do you take into account the variation in resistance with the degree of swelling of the wall?

BLACKSHEAR: The resistance of the interstitium we derived from the value of k we obtained in the steady state; we simply used Darcy's law with constants evaluated from values for steady state.

CARO: I do not agree with your statement that a necessary condition for swelling of the sub-endothelium is that luminal pressure exceeds sub-endothelial pressure. All that is necessary, I think, is to make one part of the wall more permeable than another. I can conceive a part of the wall swelling if I make the endothelium more leaky and there is restricted outflow of water via the media of the adventitia.

BLACKSHEAR: Do a free body diagram of the blister. Before it can swell the pressure has to be elevated.

ANONYMOUS: The question that I have is what then do you propose as the sequal to the blistering that is only part of the process? How does that couple into the atherogenic process subsequently?

BLACKSHEAR: Until this meeting when the gospel according to Glagov really came home to me, I would say that all one needs is to have sub-endothelial swelling occur, so as to lift the endothelium off the sub-endothelium or wherever it is tethered in the human such that you put the same type of strain that Dr. Nerem applies when he exceeds 50 dynes per square centimeter of shear stress. If we have a lifting force that is 50 dynes per square centimeter lifting the endothelium up, why doesn't that strain the endothelial gaps just as much as a shear stress? On the other hand, once the blister gets beyond a certain length it is possible for physiological shear stresses, no matter how low they are, to tear the blister off.

ANONYMOUS: Do you have any idea why they are focal?

BLACKSHEAR: If everything were completely homogeneous, if there were no disparities in the media and no pressure ripples in the lumen, then it would be a chance affair that some place would blister first. It then would drain its surrounding field, and there would be a probability of somewhere else blistering. I think it is easy in flowing blood to see that there are always going to be pressure minima. If the fluid can communicate below the endothelium, this siting effect would give the highest probability, all things being equal, to the local minimum pressure in the lumen.

WEINBAUM: It would be really interesting, if those places where the blisters are occurring, are the places where the junctions are imcomplete. Tomorrow I will show you the solutions for the experiment which you described today, because we are interested in just the question that was asked. We wanted to see if we could develop a model which would predict the local flow field pressure distribution under the endothelial space. The hypothesis which was just raised is the explanation of the phenomenon; because if you have regional variations in the inter-cellular cleft, what happens is that you get local changes in sub-endothelial pressure which will make some areas more vulnerable to this than others.

Influence of Convection and Vasoactive Agents on the Mass Transport Properties of the Arterial Wall

C. G. Caro, M. J. Lever, A. Baldwin, and A. Tedgui

Introduction

A characteristic of the complex process of the development of atheroma is the accumulation of material in the arterial wall. Our initial work in this field involved determining whether the distribution of early atheromatous lesions in arteries is correlated with the local haemodynamics. Our observations indicated that early lesions in man occur preferentially in regions where the wall shear is expected to be low (Caro et al. 1971). Support for that view has come from other laboratories (Friedman et al. 1981; Kjaernes et al. 1981). It was proposed (Caro et al. 1971) that the correlation results from dependence of transitional mass transport on wall shear (Fry 1969; Caro and Nerem 1973) and that accumulation of material in the wall is in part due to alteration of material entering the wall and inadequate efflux of material from wall to blood. Since endothelial damage markedly increases intimal permeability, the hypothesis gains support from the observation that no endothelial damage is evident in arteries (Zarins et al. 1980; Schwartz 1980). The hypothesis is also supported by the finding that low density lipoprotein is present in high concentration in arterial intimal interstitial fluid (Smith and Staples 1982).

Most investigations in this field have been concerned with the intima and intimal mass transport and take no account of possible contributions to the development of atheroma by the mass transport of the remainder of the arterial wall. Our work in the past few years with relatively inert extracellular tracers in arteries perfused in situ or incubated in vitro (Caro et al. 1980a) provides support for the long-held view that movement occurs of material right across the arterial wall from blood to adventitial lymph. It indicates, in addition, that the media represents an important transport resistance.

There is both pathological and physiological evidence consistent with intramural mass transport playing a role in atherogenesis. The former evidence includes the predilection of atheroma for thicker-walled arteries, the sparing of thinner-walled vessels and the tendency for lesions to develop between the endothelium and inner media. The physiological evidence includes the restricted mobility of macromolecules in connective tissue (Comper and Laurent 1978), the exclusion of macromolecules from arterial medial interstitium (Caro et al. 1980a) and specific interactions of lipoproteins with components of the interstitium (Bihari-Varga and Vegh 1967). Additional complexity is revealed by some recent observations from this laboratory. External forces, including arterial blood pressure and arterial smooth muscle tone, are found to alter the exclusion of plasma components from the media (Caro et al. 1980b; Caro and Lever 1982). Moreover, connective-, or pressure-driven-, flow across the arterial wall, which is influenced by factors including transmural pressure and endothelial permeability, is found to affect the uptake of tracers by the wall and seemingly the conformation of the interstitium (Lever and Tedgui 1981; Baldwin et al. 1982).

Fluid Dynamics as a Localizing Factor for Atherosclerosis
Editors: G. Schettler et al.

We report in this communication studies we have undertaken on the uptake of relatively inert extracellular tracers by excised arteries, considering in particular the influence of transmural convection and vasoactive agents. Radioactively labelled albumin was used as a high molecular weight tracer and radioactively labelled sucrose or EDTA as low molecular weight tracers. Two series of experiments were undertaken. In the first series, arteries were non-pressurized or pressurized with air to obviate transmural convection. In the second series, the vessels were liquid-pressurized, so that transmural convection would be present.

Methods

Most of the experimental procedures have been described in detail (Caro et al. 1980a). 4-7 month old New Zealand white rabbits, fed a normal diet were anaesthetized with sodium pentobarbitone and heparinized. The common carotid arteries or aorta were exposed and excised. Experiments were conducted with the vessels incubated at 38 $\pm$ 1°C when at relaxed length and zero transmural pressure (non-pressurized); at in vivo length and pressurized to 70 mmHg with air (air-pressurized) or at in vivo length and pressurized to 70 mmHg with liquid (liquid-pressurized).

The non-pressurized vessels were cut into segments about 0.5 cm in length and immersed in Tyrodes solution containing 4% bovine serum albumin together with tracer. Bubbling 5% CO_2 in air through the incubating solution did not affect tracer uptake. The air-pressurized vessels were cannulated in situ, excised and incubated without being allowed to shorten, to lessen the likelihood of endothelial damage. The downstream cannula was clamped to prevent air flow along the vessels. The external solution to simulate interstitial fluid, was Tyrodes solution containing tracer and 2% albumin. The liquid-pressurized vessels were similarly prevented from shortening. They were pressurized with Tyrodes solution containing 4% albumin and tracer, the downstream cannula again being clamped. The external incubating solution was Tyrodes solution containing 2% albumin and tracer at the same concentration as in the lumen. Smooth muscle tone was varied in some experiments with the use of noradrenaline (NA), sodium nitrite ($NaNO_2$) or isosorbide dinitrate (ISDN). In some studies all segments, including the controls, were incubated with 10^{-7}M NA, in order to induce additional tone in the excised vessels. In other studies, the endothelium was removed by passing a rubber tube along the lumen.

Following incubation, the vessels were opened axially, frozen and subjected to sequential frozen sectioning parallel to the intima at 20 µm intervals. The tissue/solution radioactivity ratio, C_T/C_P (cpm^{-3} tissue/cpm cm^{-3} incubating solution) was determined for individual sections and wall activity profiles were constructed. To permit comparison of the profiles from arteries of different animals or from arteries from the same animal whose thickness was altered experimentally, medial thickness was made equal to 100 units. Because of the relative thickness of the sections and the distinction between media and adventitia being based on the appearance of the sections, there was uncertainty as to the exact location of the medial/adventitial boundary. To overcome this problem, mean medial and adventitial C_T/C_P values were calculated by averaging values for sections lying between 0-80 and 120-180 units respectively from the lumen. In view of the long times required to reach steady values, it was presumed that no significant change occurred of wall tracer distribution in the few minutes required to prepare segments for freezing.

Results

A. Carotid Arteries

(i) Non-Pressurized Vessels. The mean medial C_T/C_P for ^{125}I albumin became steady by 90 min in vessels incubated with 10^{-7}NA, 10^{-7}M NA + 10^{-3}NaNO$_2$ and 10^{-7}M NA + 10^{-3}M ISDN. The respective values were 0.048 SD 0.008 (n = 12), 0.058 SD 0.009 (n = 12) and 0.066 SD 0.010 (n = 12). The NA value was significantly lower than the NA + NaNO$_2$ ($p < 0.02$) and NA + ISDN ($p < 0.001$) values. We have previously reported that in segments incubated without a vasoactive agent, a gradual increase of mean albumin uptake occurs after 90 min and that NA at a concentration exceeding 10^{-9}M decreases and NaNO$_2$ at a concentration exceeding 10^{-4}M increases the value about that of the controls (Caro et al. 1980b).

(ii) Air-Pressurized Vessels. The 90 min mean medial values for albumin for vessels incubated with no vasoactive agent, 10^{-5}M NA and 10^{-4}M NaNO$_2$ were respectively 0.035 SEM 0.007 (n = 9), 0.027 SEM 0.003 (n = 6) and 0.051 SEM 0.008 (n = 6). The NA values are significantly lower than the controls and the NaNO$_2$ values are significantly higher.

(iii) Liquid-Pressurized Vessels. The 90 min mean medial values for albumin for segments incubated with 10^{-7}M NA, 10^{-7}M NA + 10^{-3}M NaNO$_2$ and 10^{-7}M NA + 10^{-3}M ISDN were respectively 0.062 SD 0.011 (n = 17), 0.046 SD 0.011 (n = 11) and 0.045 SD 0.007 (n = 9). The NA values are significantly higher than those obtained with NA plus either vasodilator ($p < 0.05$ and $p < 0.001$ respectively).

B. Aorta

(i) Liquid-Pressurized Vessels. The 90 min mean medial value for albumin was 0.15 SD 0.03 (n = 4) in the de-endothelialized vessels and 0.056 SD 0.003 (n = 5) in intact vessels. The steady mean medial value for ^{14}C sucrose was 0.443 SD 0.029 (n = 7) for de-endothelialized vessels and 0.418 SD 0.015 (n = 11) for intact vessels ($p < 0.05$).

Discussion

It could be expected that transmural convection would be absent when transmural pressure was zero or vessels were pressurized with air. In the latter case inter-facial forces would prevent inflow to the wall. The arterial wall, however, compacts slowly when subjected to stresses in the physiological range. Transient convection could, therefore, follow the relaxation or air-pressurization of a vessel, but this presumably would have become small at the longer incubation times, which probably exceeded consolidation times (Blackshear et al. 1980). Convection was thought to be approximately normal in the liquid-pressurized intact vessels and was found to be enhanced in the de-endothelialized vessels. The labelled albumin is thought to have been distributed principally in the interstitium and changes in tracer uptake are believed to represent mainly changes in interstitial properties. Evidence in support of that view and against entry of tracer into cells is that the steady uptake of the smaller species ^{51}Cr EDTA was always within the normal range.

A steady-state concentration can be achieved in the interstitium both in the absence of and in the presence of convection. It need not, however, take the same value in the two conditions. A feature of the studies was the increase of the steady mean medial uptake of albumin with increase of convection; the value was lower in the absence of convection (air-pressurized vessels) than in the presence of convection (liquid-pressurized vessels) and was substantially elevated when convection was enhanced (de-endothelialized vessels). A small increase was also seen in de-endothelialized vessels of the steady mean medial uptake of ^{14}C-sucrose.

The mechanisms underlying these changes are not clearly established; however, the following considerations may help explain them. The steady-state distribution of an 'inert' extracellular tracer in the arterial wall will depend firstly on wall structure. Pressurization or axial extension of a vessel, or alteration of the state of contraction of the smooth muscle in the wall, can change wall morphology, including that of the interstitium. The structure of the wall may also be influenced by the pressure and flow of liquid within it. Consistent with these proposals is the observed reduction of the steady-state medial uptake (distribution volume) for labelled albumin with increase of luminal pressure in air-pressurized arteries (Caro et al. 1980b). Also consistent is the elevation of the steady medial uptake of both labelled albumin and sucrose in de-endothelialized liquid-pressurized arteries. It would seem from these results that distension of vessels causes compaction of the wall tissue with loss of water and concentration of components of the interstitium. Flow of water through the wall in liquid-pressurized arteries apparently acts to oppose this compaction. Convection may also influence the distribution of tracer in the wall by other mechanisms. It may influence tracer movement (for example, solvent drag which will affect less mobile materials such as macromolecules more than smaller materials). It may also affect tracer interaction with components of the wall, leading possibly to solute retention.

It is of interest to see whether these ideas aid understanding of other responses of the arterial wall. Noradrenaline reduced the diameter of the non-pressurized, air-pressurized and liquid-pressurized vessels. Its effect on tracer uptake was not, however, uniform in these experiments. It caused a reduction of the mean steady-state medial uptake of labelled albumin in the absence of transmural convection, but enhancement of that uptake when convection was present. A possible explanation for these results is that noradrenaline increased the permeability of the endothelium (as reported by Robertson and Khairallah 1972) including that to water - an action which might be opposed by vasodilators. Consistent with this idea, there was patchy staining of the intima with Trypan blue and marked convolution of a few endothelial cells in arteries exposed to noradrenaline, but not in arteries treated with noradrenaline plus a vasodilator.

Our investigations of transmural mass transport in arteries are still at an early stage. The endothelium is recognized to offer an important transport resistance. Our results indicate that the media, too, offers a significant resistance, which can be altered by factors including arterial blood pressure, transmural convection and medial smooth muscle tone. Furthermore, the transport properties of the intima and media appear to interact. For example, change of endothelial permeability, by influencing convection, appears capable of affecting the structure and mass transport properties of the interstitium of the media.

It remains to be determined what role transmural mass transport plays in the development of atheroma. It would seem possible that factors, including those mentioned above, affect the composition of the interstitium and, hence, the environment of the cells of the arterial wall. Moreover, the accumulation of material in the wall may be influenced by the magnitude of the intimal and medial resistances. Accumulation would be expected to be slight when medial transport resistance was low and wall shear might then have little effect on local accumulation.

Acknowledgements. We acknowledge helpful discussions with Professor J.T. Stuart and Drs. K.H. Parker and C.P. Winlove. Support for the work derived from the Medical Research Council, National Heart Research Fund and Sanol Schwarz Pharmaceuticals Ltd., U.K. The ISDN was kindly provided by Pharma Schwarz, Monheim, Germany.

References

Baldwin A, Lever MJ, Caro CG (1982) Effect of noradrenaline sodium nitrite and isosorbide dinitrate in albumin transport in the wall of the excised rabbit common carotid artery. Atherosclerosis (in press)

Bihari-Varga M, Vegh M (1967) Quantitative studies on the complexes formed between aortic mucopolysaccharide and serum lipoproteins. Biochim Biophys Acta 144: 202

Blackshear PL Jr, Vargas FF, Emerson PF, Newell MJ, Vargas CB, Blackshear GL (1980) Water and ion flux through the artery wall. In: Nerem RM, Guyton JR (eds) Hemodynamics and the arterial wall. Proceedings from a Specialists' Meeting, University of Houston, Texas

Caro CG, Fitz-Gerald JM, Schroter RC (1971) Atheroma and arterial wall shear. Observation, correlation and proposal of a shear dependent mass transfer mechanism for atherogenesis. Proc R Soc Lond B 177: 109

Caro CG, Lever MJ, Laver-Rudich Z, Meyer F, Liron N, Ebel W, Parker KH, Winlove CP (1980a) Net albumin transport across the wall of the rabbit common carotid artery perfused in situ. Atherosclerosis 37: 497

Caro CG, Lever MJ, Tedgui A (1980b) Effect of active and passive stresses on distribution volume in the arterial wall. In: Nerem RM, Guyton JH (eds) Hemodynamics and the arterial wall. Proceedings from a Specialists' Meeting, University of Houston, Texas

Caro CG, Lever MJ (1982) Effect of vasoactive agents and applied stress on the albumin space of excised rabbit carotid arteries. Atherosclerosis (in press)

Caro CG, Nerem RM (1973) Transport of ^{14}C cholesterol between serum and wall in perfused dog common carotid artery. Circ Res 32: 187

Comper WA, Laurent TC (1978) Physiological function of connective tissue polysaccharides. Physiol Rev 58: 255

Friedman MH, Hutchins MG, Bargeron CB, Deters OJ, Mark FF (1981) Correlation between intimal thickness and fluid shear in human arteries. Atherosclerosis 39: 425

Fry DL (1969) Certain chemorheologic considerations regarding the blood vascular interface with particular reference to coronary artery disease. Circulation 39: Suppl 4, 38

Kenyon DE (1979) A mathematical model of water flux through aortic tissue. Bull Math Biol 41: 79

Kjaernes M, Svindland A, Walldøe L, Witte SO (1981) Localisation of early atherosclerotic lesions in an arterial bifurcation in humans. Acta Path Microbiol Scand A 89: 35

Lever MJ, Tedgui A (1981) Oedema and albumin space in the rabbit aorta following intimal damage. J Physiol 319: 37

Robertson AL, Khairallah PA (1972) Effects of angiotensin II and some analogues on vascular permeability in the rabbit. Circ Res 31: 923

Schwartz SM (1980) Role of endothelial integrity in atherosclerosis. In: Nerem RM, Guyton JH (eds) Hemodynamics and the arterial wall. Proceedings from a Specialists' Meeting, University of Houston, Texas

Smith EB, Staples EM (1982) Intimal and medial plasma protein concentrations and endothelial function. Atherosclerosis 41: 295

Zarins CK, Taylor KF, Bamberger RA, Glagov S (1980) Endothelial integrity at aortic ostial flow dividers. Scan Electron Microsc 3: 249

Discussion

WEINBAUM: I want to be sure I understand the difference in the two experiments. The one where you are getting a measurement of roughly four per cent available volume for albumin, are you labelling both from the inside and outside, or just from the inside?

CARO: The data were obtained by air pressurizing arteries and incubating them with iodinated albumin applied to the outside only. We are following the time course of the wall label concentration, and we find a steady value to be achieved in about 90 minutes.

WEINBAUM: But you do not fill up the total available volume when you label from only one side, because you have a gradient going in.

CARO: I do not agree at all. We air pressurize vessels to obviate transmural convection (which occurs if the vessels are liquid pressurized), and hence we expect the equilibrium will be approached. The wall is very thin, about 150 μm, and we determine the time course of the tracer concentration in 20 μm sections cut parallel to the intima. The resolution is poor because of the thickness of the sections, but at 90 minutes a steady value is achieved as I showed. We assumed that there is no efflux at the air:wall interface and that the wall saturates with label even though it is only applied to the outside.

WEINBAUM: When do you get 20% available volume for albumin?

CARO: This is obtained when we liquid pressurized arteries and pass a soft rubber tube along the lumen to strip off the endothelium. In this case we get a 20% albumin space, and as expected, a modest (5 - 10%) increase in the available space for the smaller material, labelled sucrose. We attribute these changes to an increase in transmural convection which results from endothelial damage in the liquid-pressurized arteries.

SCHWARTZ: I was very interested in the data on the changes in available space in the wall in response to nitrates and epinephrine, because both of these have effects which appear to be mediated by the endothelium. Nitrate in particular can elicit prostaglandin release from endothelium. Recently, there is a paper which describes the acetylcholine effect on smooth muscle cells as depending entirely on the presence of endothelium. Acetylcholine is a vaso-constrictor without endothelium and a vaso-dilator with endothelium. So it would be interesting as a test of what you are saying to see if you get reverse effects with and without the endothelium present.

CARO: We would like very much to test that idea. We also have other reasons for carrying out experiments with de-endothelialized arteries. Among these is that vasoactive materials may affect the permeability of the arterial endothelium and hence transmural convection.

Haemodynamic Influences on Platelets in Haemostasis and Thrombosis

G. V. R. Born

When blood vessels are unjured so that they bleed, circulating platelets adhere to the damaged vessel walls and aggregate, so diminishing or arresting the haemorrhage. This interaction between platelets and vessel walls therefore has an easily demonstrable physiological function. There is much clinical and experimental evidence that a deficiency or defect in circulating platelets is associated with "spontaneous" haemorrhages from small vessels. This suggests that platelets are somehow essential for the functional integrity of these vessels, but no mechanism has yet been established.

Claims are made that agents released from platelets are able to damage vessel walls, either acutely (Mustard et al. 1977b) or by contributing to atherogenesis (Ross and Glomset 1973). The evidence for these propositions is indirect and cirumstantial and no such effects have been incontrovertibly established (see Walton 1975). On the other hand, there is conclusive evidence that occlusive thrombi in arteries damaged by atherosclerosis contain platelets as a major, if not the main, component (Davies and Thomas 1981). The formation of platelet thrombi appears so similar to that of haemostasis plugs of platelets that analysis of the mechanism of the latter is likely to provide an understanding of the former. This introduction poses questions about how the plugging mechanism depends on the haemodynamic environment in which platelets aggregate on vessel walls.

Both the gross and the histological appearance of arterial thrombi establish that the central mass consists mainly of aggregated platelets. What, therefore, is the mechanism responsible for rapid and extensive platelet aggregation in an artery as an apparently random event in time (see Born 1979)? Close serial sectioning of obstructed coronary arteries established some time ago that the platelet thrombus responsible is invariably associated with recent haemorrhage into an underlying atherosclerotic plaque (Friedman 1970; Davies and Thomas 1981). The haemorrhages occur through fissures or fractures in the plaque; and the sudden appearance of such a fissure or fracture may well be the random, individually unpredictable event affecting coronary arteries that has to be assumed to occur to account for the clinical onset of acute coronary thrombosis (Born 1979).

How does haemorrhage into a ruptured plaque start off platelet thrombogenesis? This can be regarded as part of the general question of how platelets are caused to aggregate through haemorrhage, and most effectively through haemorrhage from arteries. Until recently this question was commonly answered by assuming that the process depends on the adhesion of platelets to collagen which is exposed where damaged vessel walls are denuded of endothelium. Adhering platelets then release other agents, including thromboxane A_2 and ADP, which in turn are responsible for the adhesion of more platelets as growing aggregates. This explanation is unlikely to be correct, for the following reasons. First, haemostatic and thrombotic aggregates of platelets

Fluid Dynamics as a Localizing Factor for Atherosclerosis
Editors: G. Schettler et al.

grow without delay and very rapidly. For example, when an arteriole 200 µm in diameter is cut into laterally, the rate of accession of platelets to the haemostatic plug is of the order 10^4/s (Born and Richardson 1979). In contrast, although the adhesion of platelets to collagen itself is almost instantaneous, the subsequent aggregation of platelets, even under optimal conditions for their reactivity, begins only after a delay or lag period of at least 15 to 30 s (Wilner et al. 1969). Secondly, platelets tend to aggregate as mural thrombi when anti-coagulated blood flows through the plastic vessels of artificial organs such as oxygenators or dialysers (Richardson et al. 1976) that contain no collagen or anything else capable of activating platelets similarly. This implies that there are conditions under which platelets are activated in the blood by something other than collagen or other constituents of the walls of living vessels.

The plaque on which a thrombus grows has usually narrowed the arterial lumen. At constant blood pressure the flow of blood is faster through the constriction than elsewhere in the artery. Therefore, high flow and wall shear rates are no hindrance to the aggregation of platelets as thrombi (Born 1977). Indeed, the question arises of whether the activation of platelets which precedes their aggregation depends in some way on such abnormal haemodynamic conditions.

The effectiveness of platelet aggregation in plugging a leak is at least as effective in arterioles as in venules. As the haemodynamic situation should be more unfavourable to the formation of aggregation in arterioles than in venules, an explanation of arteriolar haemostasis is likely to account in principle also for that in venules. For that reason, the following considerations are limited to arterioles.

When an arteriole is cut, platelets are seen to adhere with great rapidity to the damaged vessel wall, while the red cells continue to rush by. This high flow velocity in relation to the small size of the vessels implies the presence in the fluid of strong mechanical forces acting normally and tangentially on and near the vessel walls. The cut causes peripheral resistance to the flow to diminish suddenly; and if the inflow pressure remains constant the mean flow velocity increases. Thus the fluid-mechanical forces on platelets adhering and aggregating on the vessel wall become greater still. With increasing size the platelet aggregates tend to constrict the cut, causing a further, although usually temporary, increase in flow velocity.

In spite of wall shear stresses of 10^5 to 10^6 µN/cm^2 which are one or two orders of magnitude greater than anywhere in the normal circulation, platelets succeed into haemostatically effective plugs. The blood-flow velocities that would be experienced closest to the vessel wall and therefore with the highest probability of colliding with the sites of damage can be calculated (Schmid-Schönbein et al. 1976). Human platelets have a major diameter of about 1.5 µm. In an arteriole of medium size the flow velocity of plasma and of any cells in it at a distance of 1 µm from the wall is of the order of 10-100 µm/ms. Therefore, a platelet flowing within a distance no greater than its own diameter would pass an injury site 100 µm long in at most 10 ms. In the absence of other influences, this would seem to be the time available for such a platelet to adhere to the damaged wall.

The time just calculated as available to circulating platelets 'at risk' for adhering to a wall lesion has to be compared with what is known about the time required for platelets to be activated into a condition in which their collision with such a lesion would very probably result in adhesion. That a process of activation is an essential prerequisite for adhesion and aggregation is inferred from the non-reactivity of normal circulating platelets.

As activation is indicated by adhesiveness, the change must involve one or more constituents of the outer surface of platelets. There is evidence that the essence is the exposure of surface receptors for fibrinogen, which has long been known to be an essential and specific plasma co-factor for platelet aggregation (Born and Cross 1964; Cross 1964). The activation time of platelets may then be defined as the interval between the encounter of platelets with an activating agent such as ADP and their ability to react with plasma fibrinogen.

That circulating platelets can be activated to adhere in much less time than that required by their gross changes in shape (Born 1970) is indicated by direct experimental observations. An arteriole can be irradiated by a laser in such a way that damage is limited to a few square micrometers of endothelium (Arfors et al 1976). The site of damage is covered almost immediately with platelets that must have been activated in small fractions of a second.

Very similar events follow the application of the activating agent ADP by micro-iontophoresis to the outside of an arteriole or venule under conditions in which appropriate controls indicate that there is no evidence at all of damage to the endothelial layer (Begent and Born 1970). Platelet aggregates grow in the vessel exactly opposite the tip of the micropipette, while the blood continues to flow rapidly and without noticeable disturbance over the site. This is explained most simply by assuming that sufficient ADP diffuses between the endothelial cells into the blood to reach platelets passing close to the wall and that this ADP activates them in a few seconds.

An extension of this technique has provided a basis for calculating an average activation time for circulating platelets. It was found that the size of platelet aggregates produced by the iontophoretic application of ADP increases exponentially. The rate constant of this increase depended on the mean blood flow velocity, determined in the same vessels at the same time (Begent and Born 1970). The shape of the experimentally determined curve was simulated closely by a theoretical curve (Richardson 1973), which was derived on the single assumption that platelets require an activation time of about 100 ms to 200 ms. This time is still one order of magnitude greater than that indicated by the earlier theoretical considerations, so either this experimental derivation overestimates the true activation time or the earlier considerations failed to take something into account that would allow flowing platelets more than a few milliseconds for activation. More time would, for example, be available if the blood flow near the vessel wall were non-laminar, so that platelets caught up in vortices, however small, might be exposed to localized activating conditions for longer than they would otherwise be. When branching vessels of the microcirculation are observed microscopically, platelets can often be seen trapped in vortices for variable times of up to several seconds. Such delays may occur in the immediate vicinity of major vessel wall lesions, whether caused by disease such as the sudden rupture of an atheromatous plaque (Friedman and Van den Bovenkamp 1966; Davies and Thomas 1981) or by traumatic injury such as a puncture or transection. However, there is no evidence of even the smallest disturbances in the flow of blood in a normal vessel in which platelets are caused to adhere by iontophoretically applied ADP. Moreover, it seems most unlikely that any endothelial uneveness produced by laser injury would give rise to flow disturbances large enough to delay the passage of platelets.

The old thrombogenic hypothesis of atherosclerosis (von Rokitansky 1841; Duguid 1949) has recently reappeared in modern costume as claims that *platelets* contribute to atherogenesis in three ways: first, through damaging arterial endothelial cells by releasing injurious agents, presumably where circulating platelets adhere (Mustard et al. 1977a); secondly, through the release in such situations of a factor responsible for smooth muscle proliferation in the arterial wall (Ross and Glomset 1973); and thirdly through the formation of persistent mural thrombi which are organized into intimal thickenings. Such evidence as there is for these propositions fails to establish any of them as relevant to atherosclerosis in animals or human beings (see Walton 1975). Underlying all three claims is the assumption that some normal circulating platelets settle on arterial walls for long enough to release some of their contents. There is no observational basis for this assumption in normal arteries. Therefore it is assumed further that arterial endothelium is continuously subject to 'damage' or 'injury' of some kind as a precondition for the adherence of platelets. There is no convincing evidence for this generalization, especially not in human beings. The only finding that could conceivably apply to human arteries is that guinea-pig aorta has a higher replacement rate of endothelium around the openings of branches than elsewhere (Payling-Wright and Born 1971). This is most simply explained by assuming that endothelial turnover depends, *inter alia* , on haemodynamic effects due to non-laminar blood flow over such areas. But this should be thought of more correctly as a quasi-physiological effect and, even there, platelets are rarely if ever seen adhering to the walls. The turnover rate of endothelium is increased in experimental hypertension (Payling-Wright 1972). This is compatible with hypertension as a 'risk factor' for coronary heart disease. It seems more likely that this is due to an accelerating effect of interendothelial gaps on plasma lipoprotein accumulation (Stehbens 1965; Caro 1977) than to an increase in the indiscriminate or even selective deposition of platelets on arterial walls.

This Session is therefore concerned inter alia with questions, most of them unanswered, about the effects of haemodynamics in the interactions of cellular and other constituents of the circulating blood with the walls of the vessels, particularly with those of arteries.

References

Arfors KE, Cockburn JS, Gross JF (1976) Measurement of growth rate of laser-induced intravascular platelet aggregation, and the influence of blood flow velocity. Microvasc Res 11: 79-87

Begent NA, Born GVR (1970) Growth rate in vivo of platelet thrombi produced by iontophoresis of ADP, as a function of mean blood flow velocity. Nature (Lond) 227: 926-930

Born GVR (1970) Observations on the change in shape of blood platelets brought about by adenosine diphosphate. J Physiol 209: 487-511

Born GVR (1977) Fluid-mechanical and biochemical interactions in haemostasis. Br Med Bull 33: 193-197

Born GVR (1979) Arterial thrombosis and its prevention. In: Proc VIII World Congr Cardiol, Tokyo 1978. Excerpta Medica, Amsterdam, ICS 470, p 81-91

Born GVR, Cross MJ (1964) Effects of inorganic ions and of plasma proteins on the aggregation of blood platelets by adenosine diphosphate. J Physiol (Lond) 170: 394-414

Born GVR, Richardson PD (1979) Activation time of blood platelets. Proc R Soc Lond (Biol) (submitted)

Caro CG (1977) Mechanical factors in atherogenesis. In: Cardiovascular flow dynamics and measurement. University Park Press, Baltimore

Constantinides P (1966) Plaque fissures in human coronary thrombosis. J Atheroscler Res 6: 1-17
Cross MJ (1964) Effect of fibrinogen on the aggregation of platelets by adenosine diphosphate. Thromb Diath Haemorrh 12: 524-527
Davies MJ, Thomas T (1981) The pathological basis and microanatomy of occlusive thrombus formation in human coronary arteries. Phil Trans R Soc Lond B 294: 225-229
Duguid JB (1949) Pathogenesis of atherosclerosis. Lancet 2: 925
Friedman H (1970) Pathogenesis of coronary thrombosis, intramural and intraluminal haemorrhage. In: Halonen LA (ed) Thrombosis and coronary heart disease. Karger, Basel, vol 4, p 3
Friedman M, Van den Bovenkamp GJ (1966) The pathogenesis of a coronary thrombosis. Am J Pathol 48: 19-44
Mustard JF, Moore S, Packham MA, Kinlough Rathbone RL (1977a) Platelets, thrombosis and atherosclerosis. Proc Biochem Pharmacol 13: 312-325
Mustard JF, Packham MA, Kinlough Rathbone RL (1977b) Platelets, thrombosis and atherosclerosis. Adv Exp Med Biol 104: 127-144
Payling-Wright HP (1972) Mitosis patterns in aortic endothelium. Atherosclerosis 15: 93-95
Payling-Wright HP, Born GVR (1971) Possible effect of blood flow on the turnover rate of vascular endothelial cells. In: Hartert HH, Copley AL (eds) Theoretical and clinical hemorheology. Springer, Berlin Heidelberg New York, p 220-226
Richardson PD (1973) Effect of blood flow velocity on growth rate of platelet thrombi. Nature (Lond) 245: 103-104
Richardson PD, Galetti P, Born GVR (1976) Regional administration of drugs to control thrombosis in artificial organs. Trans Am Soc Artif Intern Organs 22: 22-29
Ross R, Glomset JA (1973) Atherosclerosis and the arterial smooth muscle cell. Proliferation of smooth muscle is a key event in the genesis of the lesions of atherosclerosis. Science (Wash) 180: 1332-1339
Schmid-Schönbein H, Rieger H, Fischer T (1976) Blood vessels: problems arising at the borders of natural and artificial blood vessels. In: Effert S, Meyer-Erkelenz J (eds). Springer, Berlin Heidelberg New York, p 57-63
Stehbens WE (1965) Endothelial cell mitosis and permeability. J Exp Physiol (Cogn Med Sci) 50: 90-92
von Rokitansky (1841-46) Handbuch der pathologischen Anatomie. Braumüller & Seidel, Vienna
Walton KW (1975) Pathogenetic mechanisms in atherosclerosis. Am J Cardiol 35: 542-558
Wilner GD, Nossel HL, LeRoy EC (1969) Aggregation of platelets by collagen. J Clin Invest 47: 2616-2621

Discussion

WEINBAUM: I was wondering whether you have a hypothesis to explain the permeability data you obtained for the neuraminidase? I noticed from your albumin data that there is not much change, so I would assume since the albumin is probably passing through the cleft, that the cleft dimensions have not changed singificantly. Is the LDL going through a different pathway than the cleft? What do you think about that?

BORN: I agree with you. The question is where does the LDL go through? The absolute rates of movement of albumin are of course much higher than those for LDL, as you would expect.

S. SCHWARTZ: One question I have is about the experiment where you iontophorese vasoactive agents near vessels. The endothelial cells are pharmacologically active, they will presumedly respond to histamine, for example, by releasing PGI2 and presumedly PGE2 and thromboxane.

How can you be sure that you are looking at a direct response as opposed to secretion of pharmacological activations?

BORN: In the early experiments we could not because it was before thromboxane and prostaglandin. Now we are doing this, and I can tell you already that, if both are abolished with indometacin, the results are the same. Evidence with a thromboxane synthesis inhibitor is that again there is no change. You may be interested to know that Colin Dolery in London, with the most refined techniques, finds no prostacyclin at all in the circulating human blood, i.e. the values are not significantly different from zero. This does not mean that when these tissues contract they might not release some prostacyclin, but the amount would probably be miniscule. At present the evidence is against it doing anything.

CARO: You perfuse for a period of several hours. Is that right?

BORN: We perfuse for half an hour.

CARO: My question is whether perfusion with saline per se has any long-standing effect on permeability. I believe that Charles Michel working with capillaries finds that it does, and so does Peter Winlove in my laboratory working with arteries.

BORN: To take account of Charles Michel's observations, our perfusion medium is Krebs solution with an adequate concentration of plasma albumin.

C. SCHWARTZ: In measuring the amount of albumin present in the vessel after perfusion with isotopic albumin, did you indeed at that point wash or use saline in the manipulation? It is just remotely possible that this could be a key issue here because of the ionic removal of bound isotopic albumin as part of that, and this could partly account for the non-difference perhaps in this sort of a situation. I do not know.

BORN: That is unlikely because the same amount is on the walls on both sides, whether washed or unwashed. So if you wash away whatever is absorbed, much has gone into the wall.

Effects of Antihypertensive Drugs and Blood Velocity [1]

J. D. Spence

The complexity of pulsatile flow at bifurcations has posed great difficulties in establishing precisely which patterns of flow, or which flow disturbances may be most important in the pathogenesis of atherosclerosis. It seems likely that different kinds of flow disturbance may be important in the progression of disease at sites of stenosis, than in the genesis of early atherosclerotic lesions. This paper takes one step back from the microrheology of atherogenesis to the hemodynamic properties such as heart rate and blood velocity, that are likely to determine the occurrence of a variety of flow patterns and flow disturbances.

A line of investigation is presented, pertaining to the hypothesis that antihypertensive drugs which minimize the occurrence of arterial flow disturbances, may be more effective in preventing the development and progression of atherosclerosis, than drugs which aggravate arterial flow disturbances.

Our studies began in Rhesus monkeys, made hypertensive with continuous infusion of angiotensin and isoproterenol [1]. It was shown that antihypertensive drugs have markedly different effects on blood velocity. In doses which reduced mean blood pressure by 15 to 20%, hydralazine increased mean peak velocity by 52%, while propranolol reduced mean peak velocity by 17%. Methyldopa and clonidine had less pronounced effects. Following the monkey studies, it was confirmed in man, using Doppler aortic velocity methods developed by Boughner [2], that antihypertensive drugs also had different effects in man [3]. In a crossover study, 15 hypertensive patients received a sequence of placebo, hydrochlorothiazide, and then hydrochlorothiazide plus nadolol, propranolol, or hydralazine. Compared to placebo, thiazide had only a slight effect on blood pressure, and little effect on heart rate or blood velocity. The beta-blockers, as would be expected, caused a marked reduction in heart rate, with a marked increase in stroke volume index. As a result, they caused a substantial increase in peak velocity, whereas hydralazine, which increased heart rate, caused a substantial reduction in peak velocity. These results were what might be expected, but were difficult to interpret, since the drugs had opposite effects on blood velocity in monkeys.

From the work of Yellin [4], it might be expected that increased heart rate, by reducing damping time, (and perhaps by its effect on rate of change of blood velocity and rate of change of shear rates) would be a factor favouring disturbed flow. On the other hand, the work of

1 This work was supported by the Ontario Heart Foundation (3-10), the Medical Research Council of Canada (MA-7118), and by the Physicians of Ontario through the P.S.I. Foundation

Fluid Dynamics as a Localizing Factor for Atherosclerosis
Editors: G. Schettler et al.

Nerem and Seed [5] suggests that increasing heart rate makes turbulence less likely as peak velocity increases. It was difficult, therefore, to predict the effect of these hemodynamic changes on arterial flow disturbances.

In order to test the effects of antihypertensive drugs on disturbed flow, studies were done in patients with angiographically proven carotid artery stenosis [3]. A Doppler flow mapping device [6] (Echoflow, Diagnostic Electronics, Lexington, Mass.) was used. This device presents a map of arterial blood velocities, using colour coding to show the distribution of blood velocities in three ranges. Red is used to represent the normal range of arterial velocity (from 20 cm/sec to 90 cm/sec), yellow represents accelerated velocities (90 to 135 cm/sec), and blue represents velocities in the range above 135 cm/sec, thought to represent turbulence. The evidence that such Doppler shifts represent turbulence in the carotid includes calculations based on known carotid diameters, using normal values for density and viscosity of blood. The occurrence of blue on arterial flow maps is taken as evidence of stenosis, and correlates highly with significant stenosis (96% of patients with stenosis of 74% or more will have areas of blue downstream for the stenosis) [7].

Twenty patients with angiographically proven stenosis had arterial flow maps done before drug, and then following hydralazine and propranolol administered intravenously, in doses which respectively raised and lowered heart rate by ten beats per minute from baselines. Arterial flow maps were read in a batch following the study, and the three independent observers were blinded as to the drug administered. Propranolol reduced the occurrence of blue and yellow regions in 19 of 20 cases, whereas hydralazine increased the occurrence of blue and yellow flow patterns in 17 or 20 cases [3]. These results suggested that, at least in regions of carotid stenosis, propranolol would minimize the occurrence of arterial flow disturbances, whereas hydralazine would aggravate them.

In order to test the effect of these differences on the development of atherosclerosis, studies were carried out in a hypertensive cholesterol fed rabbit model. Eighty-six New Zealand White Rabbits, initially weighing 3 kilograms each, were randomly allocated to five experimental groups. Controls included normotensive, normal diet animals, normotensive cholesterol-fed animals, and hypertensive cholesterol-fed animals. The two experimental groups were hypertensive cholesterol-fed animals treated with either hydralazine or propranolol, given twice daily by subcutaneous injection. Hypertension was induced by a one-kidney Goldblatt procedure (removal of the right kidney, and clipping of the left renal artery with a clip having an internal gap of 0.6 mm as described by Brooks and Muirhead [9]). Normotensive control animals underwent sham operation with removal of the right kidney, and exposure of the left renal artery. A hypertensive baseline was established for two weeks following the Goldblatt procedure, and animals that failed to develop hypertension were excluded, as they did not conform to the model. Cholesterol feeding began at week two, and was continued for the six week experimental period, during which time the treatment groups received the antihypertensive drug.

At six weeks the animals were killed, and the aortae were removed and pinned back for fixation. Extent of surface involvement of the aorta with atherosclerosis was measured from photographs, on which the atherosclerotic lesions were identified in ink, using a dissecting microscope and concurrent study of the actual aortic specimen, to verify the photographic lesions. A Tektronix 4051 Graphics Tablet operated by a Tektronix computer was used to measure the lesions, and express them as percent of aortic surface area [10].

As expected, the normotensive, normal diet rabbits had no atherosclerosis, the cholesterol-fed normotensive animals had approximately 10% of aortic surface area involved, whereas the hypertensive cholesterol-fed animals had a mean of 24% of aortic surface area involved. Hydralazine-treated animals had no significant reduction of aortic atherosclerosis, even though they had significantly lower blood pressures than the propranolol-treated animals. The propranolol-treated rabbits had significantly less aortic atherosclerosis than either the hypertensive cholesterol-fed controls or the hydralazine-treated animals. Propranolol treatment resulted in a marked decrease in heart rate, compared to hydralazine-treated animals or hypertensive controls, suggesting that the difference in development of aortic atherosclerosis may be on a hemodynamic basis.

Since these studies were done, similar results have been obtained by Pauletto et al., at Padua, in hypertensive turkeys [11]. The possibility that these differences in the development of atherosclerotic lesions in the animals treated with propranolol versus hydralazine might be on the basis of direct effect of these drugs on arterial wall, was raised by Seed at this meeting, and must be considered. The marked differences in hemodynamic effects of these drugs, and the differences demonstrated in patients with carotid stenosis, are consistent with the hypothesis that antihypertensive drugs with different hemodynamic effects may have different effects on the development and progression of atherosclerosis. Obviously, much further study is needed, and it is in the hope that investigators interested in the hemodynamic theories of atherosclerosis will be stimulated to study this problem further, that this line of investigation is presented,

References

1. Spence JD, Pesout J, Melmon KL (1977) Effects of antihypertensive drugs on blood velocity in rhesus monkeys. Stroke 8: 589-594
2. Boughner DR (1975) Assessment of aortic insufficiency by transcutaneous Doppler ultrasound. Circulation 52: 878-879
3. Spence JD (1982) A new approach to the prevention of atherosclerosis: effects of antihypertensive drugs on flow disturbances in human carotid stenosis. In: Dr. Diethrich (ed) Noninvasive assessment of the cardiovascular system. Wright PSG International Publishers, Littleton, Mass. (in press)
4. Yellin EL (1966) Laminar-turbulent transition process in pulsatile flow. Circ Res 19: 791-804
5. Nerem RM, Seed WA, Wood NB (1972) An experimental study of the velocity distribution and transition to turbulence in the aorta. J Fluid Mech 52: 137-160
6. Curry GR, White DN (1977) Color coded ultrasonic differential velocity arterial scanner (echoflow). Ultrasound Med Biol 4: 27-35
7. White DN, Curry GR (1978) A comparison of 424 carotid bifurcations examined by angiography and the Doppler echoflow. Ultrasound Med 4: 363-376
8. Spence JD, Perkins GD, Kline RL (1982) Modification of aortic atherosclerosis in hypertensive cholesterol-fed rabbits. Proceedings of the 6th International Symposium on Atherosclerosis, Berlin, Germany, June 13-17, #456
9. Brooks B, Muirhead EE (1971) Rigid clip for standardized hypertension in the rabbit. J Appl Physiol 31: 307
10. Spence JD, Perkins DG, Kline RL, Haust MD (1982) Aortic atherosclerosis in hypertensive rabbits treated with antihypertensive agents having different effects on arterial flow disturbances. 1. Extent of surface involvement. Arterial Wall (in press)

11. Pauletto P, Pessina AC, Pagnan A, Thiene G, Semplicini A, Vescovo G, Mazzucato A, Ambroso G, Dal Palu C (1982) Effect of oxprenolol on aortic atherosclerotic lesions in hypertensive broad breasted white turkeys (BBWT). Proceedings of the 6th International Symposium on Atherosclerosis, Berlin, Germany, June 13-17, #544

Discussion

SEED: I would just like to enter a plea that you try to make some more direct measurement of the flow disturbances. Some years ago Bob Nerem and I did a series of experiments in which we tried to define the parameters that determined whether you got major flow disturbances in the aorta, and if I were to summarize these results, I would say that the way you get flow disturbances in the aorta would be to slow the heart rate and increase the peak velocity.

SPENCE: One of the obvious complicating factors in the discussion today is the question of the differences between flow disturbances that are important in the pathogenesis of the earliest lesions of atherosclerosis and the flow disturbances that may be important in the progression of plaques that are already stenosing, It seems quite clear that turbulence occurs in relation to stenosis, and it may be that these hemodynamic differences between the drugs are more important in patients that have already developed lesions than in the earliest individuals with hypertension who have not yet developed lesions.

SEED: I was referring primarily to your last experiments where you showed the aortas, where presumedly you are dealing with very early disease, and where the difference in the severity of lesions was not in accord with what I would have predicted to have been the frequency of incidence of flow disturbances.

C. SCHWARTZ: The literature shows an association between the type of anti-hypertensive agent and a number of vascular phenomena which may or may not relate to pressure including myocardial hypertrophy and vascular cell proliferation. This is the first time that I have seen evidence like this for an effect on atherosclerosis. I do not know that I would agree that it is a flow parameter effect. I have no idea whether it is or not, but I think it is quite important clinically.

SPENCE: When I presented this paper as a poster in Berlin I was approached by Dr. Pauletto from Padua who told me that they had shown similar results in hypertensive turkey aortas, where they were able to almost make endothelial thickening disappear with propranolol, whereas hydralazine actually made the endothelial thickening worse. Their findings seem, therefore, to corroborate ours. The clinical data is really very interesting. The paper by Beevers et al. in the Lancet in the treatment of already stroked individuals showed that methyldopa could prevent recurrent strokes, but not heart attacks. The VA-Cooperative Trial, which used the fixed combination including hydralazine, similarly prevented strokes, but not heart attacks. There is an exciting anomalous result from the study in Göteborg in which they used mainly propranolol. They had a 50% reduction in myocardial infarction, but no reduction in strokes. These discrepancies in the results of the hypertension treatment trials would tend to support the notion that the hemodynamic effects of drugs may be important.

General Properties of Endothelial Cells

N. Freudenberg

Introduction

This session deals predominantly with endothelium, and is to be introduced by observations on the general properties of endothelial cells. Of course, we first have to consider the question whether general properties of the vascular endothelium can be said to exist at all. Considering the endothelium from the light microscopic point of view only, there is indeed a generally accepted definition that "the endothelium provides the lining of the blood vessels, heart chambers and lymphatic vessels by forming a single layer of cells". But under the electron microscope endothelial cells show structural differences which depend upon their location in the vascular system. These ultrastructural pecularities of the endothelium led to a classification of endothelial cells into different morphological types. Further investigations considering both the morphological and functional aspects of this tissue revealed, however, that transformations between the different types of endothelial cells can occur. These observations indicate that the vascular endothelium should be regarded as a dynamic structure. The overall distribution of the endothelium within the body suggests that we are dealing with the most frequently occurring type of mammalian cell, and one which also plays a predominant role with regard to number and weight.

The "Häutchen" method (Sinapius 1952) had been standardized, and after the electron microscope and, finally endothelial cell culture (Jaffe et al. 1973) had been introduced.

Morphology

Light Microscopy

In histological sections endothelium can be particularly clearly distinguished by means of the cell nucleus. Within the region of the nucleus the endothelial cell shows a distinct protrusion into the vascular lumen. The "Häutchen" preparation, on the other hand, offers far better possibilities for observing the endothelium under the light microscope. It is possible to obtain a monolayer of endothelial cells completely free from contamination with other tissue. In the silver stained "Häutchen" preparation the cell borders are outlined by black lines. This method allows the polygonal shape of endothelial cells to be demonstrated particularly clearly. After staining the "Häutchen" preparation with hematoxylin and eosin, the longitudinal axis of endothelial cell nuclei shows an alignment in the same direction as that of the blood or the lymph flow. Changes in the direction of the flow are reflected particularly well in the endothelium near the origins of branches. Figure 1 shows endothelial cells arranged in a spiral near the origin of an intercostal artery from the aorta.

Fluid Dynamics as a Localizing Factor for Atherosclerosis
Editors: G. Schettler et al.

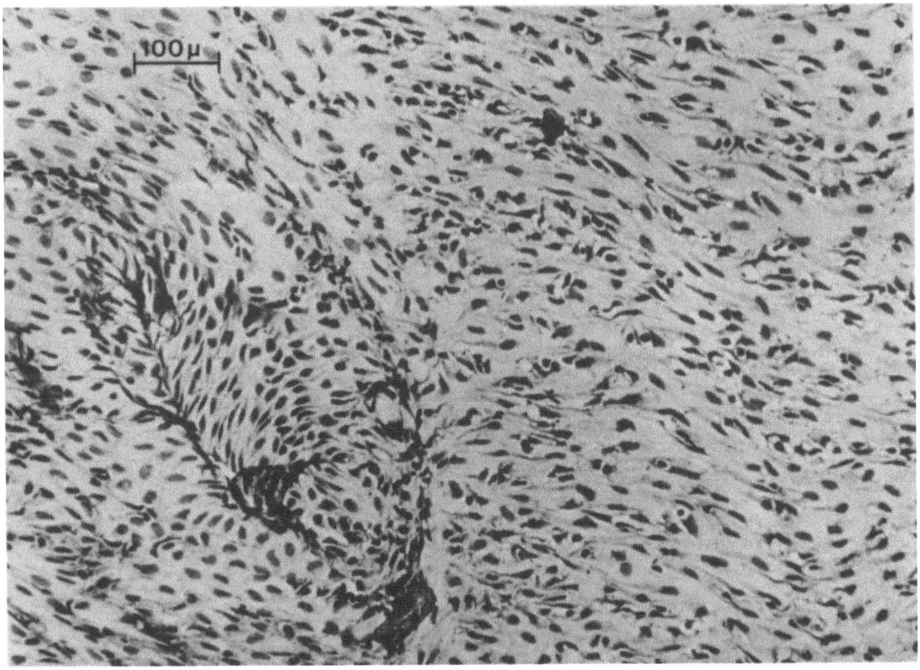

Fig. 1. Endothelial "Häutchen" of rat aorta including the origin of an intercostal artery. H.E. staining (from Friese et al. 1978)

Electron Microscopy

Under the transmission electron microscope endothelial cells show ovoid nuclei (frequently elongated) and often a large amount of cytoplasm. Hyperchromasia is frequently found near the nuclear membrane. The nuclear region of the endothelial cell shows a protrusion toward the vascular lumen. Endothelial cytoplasm is supplied with rough endoplasmic reticulum and a limited number of mitochondria with few cristae. Golgi-apparatus and centrioles are sometimes present. Glycogenic granules, ribosomes and "autolytic" vacuoles may sometimes be observed within the cytoplasm of the endothelium. The Weibel-Palade body must be considered as an organelle specific to the endothelium. It is rod-shaped and has a length of approximately 3 µm and a diameter of about 0.1 µm. The proportion of Weibel-Palade bodies compared to the total cytoplasmic volume of endothelial cells varies in different vascular regions (Steinsiepe and Weibel 1970). Micropinocytotic vesicles are characteristic structures of the endothelial cytoplasm. These electron lucent vesicles with diameter of 600-700 Å may be found throughout the cytoplasm of the endothelium. Filaments are also frequently seen in the cytoplasm of endothelial cells. In different vascular regions the arrangement of endothelial filaments ranges from densely packed bundles to loosely distributed networks. Depending upon its cytoplasmic ultrastructure three different types of endothelium may be distinguished:

1. Continuous endothelium
   This type shows a closed cytoplasm. Continuous endothelium is found in most of the large arterial, venous and lymphatic vessels.
2. Fenestrated endothelium
   This type exhibits endothelial fenestrations within extremely flattened parts of the cytoplasm. These fenestrations are either

closed by a membrane-like diaphragm or present themselves as pores without diaphragm. Both types of fenestrated endothelium possess an uninterrupted basement membrane. Fenestrated endothelium is frequently found in capillaries and particularly in the postcapillary (venous) part of the microcirculation.

3. Discontinuous endothelium
   This type of endothelium shows intercellular gaps of varying width and a discontinuous or entirely absent basement membrane. Discontinuous endothelium is characteristic of the sinusoids in the liver, spleen and bone marrow.

The surface of the luminal plasmalemma of endothelium shows two types of cytoplasmic protrusions. The first type constitutes a tentacle several µm long. After having reached a certain length these tentacles may bend toward the cytoplasmic membrane and make contact with the plasmalemma by developing a vacuole. The protrusions regularly seen near intercellular junctions are called marginal folds. Under the scanning electron microscope the marginal folds of the aortic endothelium show a ridge-like structure which marks the borders of endothelial cells distinctly. The second type of cytoplasmic protrusion is represented by endothelial microvilli seen at the luminal surface of endothelial lining. These protrusions have a maximum length of 1 µm and include neither organelles nor micropinocytotic vesicles. Microvilli are frequently found in the endothelial surface of capillaries.

Well-Established Functions

Taking into account both the light microscopic and ultrastructural appearance of endothelial morphology, current views on its known functions will be described. These functions of the vascular endothelium and their interactions are illustrated by Fig. 2. As can be seen the endothelial transport systems are closely related to metabolic activities and cellular interactions of endothelium. As a boundary lying between blood and other tissues the endothelium plays a significant role in the exchange of substances. The following endothelial structures represent the morphological counterpart to the transendothelial transport of water soluble molecules: transport channels near the intercellular junctions, micropinocytotic vesicles and fenestrations. Fig. 3 shows the metabolic activity of the endothelium as a factor maintain-

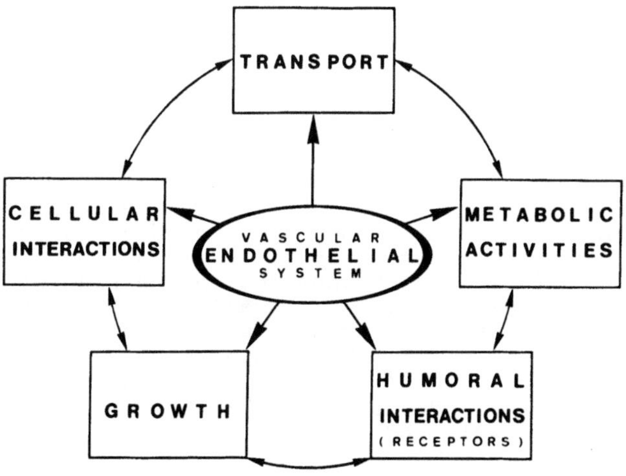

Fig. 2. Diagram showing the functions of the vascular endothelial system and their interactions (from Freudenberg et al. 1982)

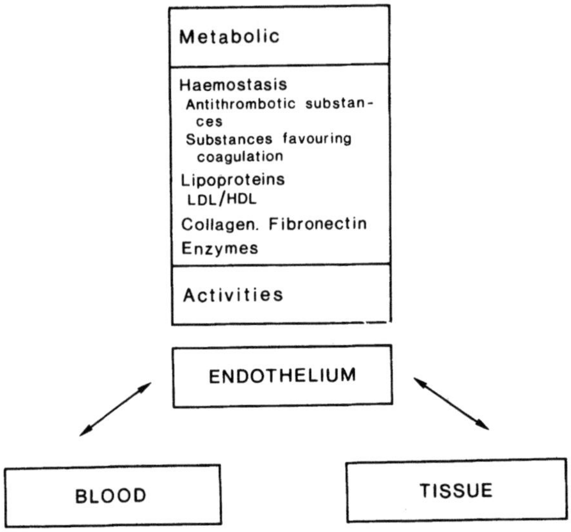

Fig. 3. Diagram demonstrating the metabolic activity of the endothelium as a factor for coordinating function between vessels and tissue (from Freudenberg et al. 1982)

ing a proper functional relationship between vessel and tissue. The most important of these endothelial functions are: the contribution to the maintainance of the hemostatic equilibrium, the metabolism of lipoproteins, the synthesis of collagen and fibronectin and enzymatic activity.

A particularly close interaction seems to exist between the metabolic activities of the endothelium and the humoral factors for which the endothelium possesses receptors. The following substances and functional groups have been identified to have receptors in endothelial cells: Vasomotor substances, clotting factors, growth factors, mucopolysaccharides, hormones and neurotransmitters, plasma lipoproteins and parts of the immune system.

While studying the proliferation processes of endothelium in our laboratory we observed that the growth behaviour of endothelial cells show distinct differences which depend on their location within the vascular system. In the large vessels the endothelium shows no homogeneous pattern of proliferation, the appearance of endothelial islands synthetizing DNA being observed (Schwartz and Benditt 1973).

Under the light microscope we were able to demonstrate all stages of mitotic cell division in the aorta and the thoracic duct. The mitosis of an endothelial cell in a small pulmonary vessel was also observed under the electron microscope. Table 1 shows a survey of the ^{3}H thymidine indices of 16 different sections of vessels from adult male Wistar rats and of two aortic sections from juvenile animals. The ^{3}H thymidine index indicates the degree of DNA synthesis and may be considered as a parameter of the proliferative activity of endothelial cells. The sections were given serial numbers (1-18) which depended on the size of the mean value of their ^{3}H thymidine indices (1 = smallest, 18 = largest mean value). The lowest DNA synthesis can be observed in endothelial cells of the distal section of inferior vena cava, whereas the most extensive DNA synthesis can be seen in the pulmonary capillaries. Based on these differences regarding the degree of DNA synthesis three main groups can be distinguished within the vascular sections of the endothelium which we examined:

Table 1. ^{3}H-thymidine indices (in %) of endothelial cells from different vascular regions in male control rats following "3-pulse labelling" (Schwartz and Benditt 1973)

Number of the vascular region	Vascular region	^{3}H-thymidine index (in %)	Significant difference to number of vascular region
1.	Vena cava inferior dist.	0.13 ± 0.05	6.,9.,11.,17.,18.
2.	Abdominal aorta of juvenile rats (2 months old)	0.14 ± 0.11	11.,14.,16.,18.
3.	Renal arteries	0.26 ± 0.15	18.
4.	Hepatic arteries	0.26 ± 0.11	18.
5.	Pulmonary veins	0.28 ± 0.26	18.
6.	Thoracic aorta	0.28 ± 0.28	1.,16.,18.
7.	Inf. vena cava (prox.)	0.30 ± 0.26	18.
8.	Abdominal aorta	0.33 ± 0.25	18.
9.	Pulmonary arteries	0.34 ± 0.15	1.,18.
10.	Thoracic aorta of juvenile rats (2 months old)	0.36 ± 0.26	18.
11.	Central hepatic veins	0.44 ± 0.17	1.,2.,18.
12.	Small pulmonary vessels	0.48 ± 0.23	1.,18.
13.	Thoracic duct	0.50 ± 0.25	1.,18.
14.	Liver veins	0.50 ± 0.21	1.,2.,18.
15.	Medullary capillaries of kidneys	0.52 ± 0.31	1.,18.
16.	Renal veins	0.88 ± 0.51	1.,2.,6.
17.	Liver sinusoids	1.12 ± 0.82	1.
18.	Pulmonary capillaries	1.14 ± 0.17	1. - 15.

1. Slightly proliferating endothelium (distal vena cava inferior)
2. Moderately proliferating endothelium (most of all investigated vessels)
3. Extensively proliferating endothelium (pulmonary capillaries)

In the second group two subgroups can be distinguished by the different standard deviations of the mean values:

- homogeneously proliferating endothelium
- heterogeneously proliferating endothelium

The varying degree of DNA synthesis in the endothelium of different vascular sections may possibly indicate that there are differences in the functional stress.

The interaction between endothelial and other cells has been known for a long time, particularly with regard to the migration of blood cells from the vascular lumen into the tissue during inflammatory processes. An occasional contact of leukocytes and platelets with the endothelium is well known. In our investigations we were able to demonstrate that in the large vessels (aorta, vena cava, thoracic duct) granulocytes, monocytes, macrophages, lymphocytes and sometimes smooth muscle cells must be regarded as belonging to the normal cell population of the intima. Little is known about the functions of these non-endothelial intimal cells or their interactions with the endothelium. It appears probable that both monocytes and macrophages play a part in the physiological process of repair in the intima.

Ultrastructural Reaction Patterns

By examining the endothelium under the electron microscope we found that endothelial cells show different structural patterns in their reactions to pathological stimuli. Depending on the quality, the intensity and the duration of a pathological stimulus, alterations indicating a degeneration and/or an activation of endothelial cells may be seen. From the type of ultrastructural alteration seen in the endothe-

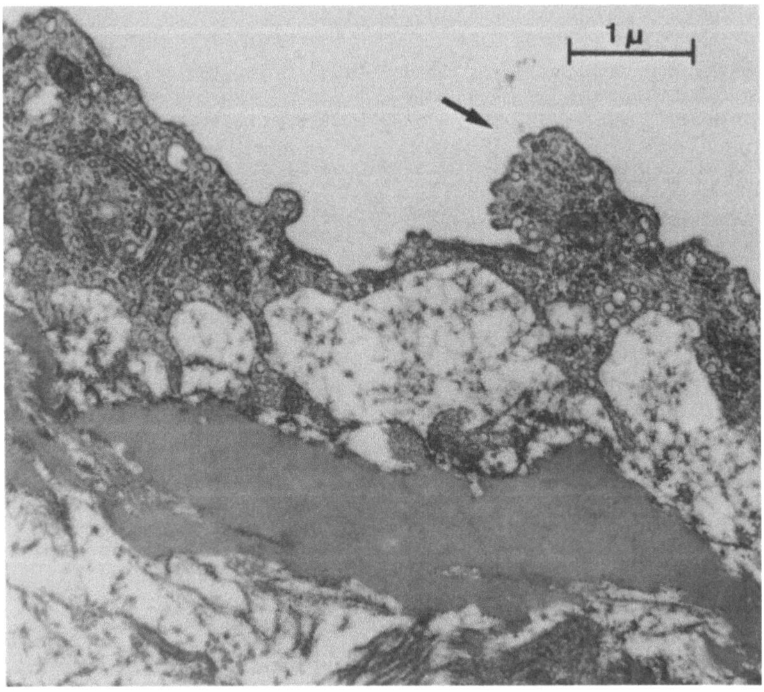

Fig. 4. Broad-based cytoplasmic extrusion (→) of the luminal endothelial surface (so-called "bleb") indicating early endothelial damage. Rat aortic andothelium, 6 hours after endotoxin injection (from Freudenberg et al. 1982)

lium we were also able to draw certain conclusions about whether acute or chronic damage had occurred. In the case of acute injuries distinction between intense or mild pathological stimuli is not clearly reflected in the ultrastructure of the individual endothelial cell. Comparable ultrastructural changes indicating degeneration are brought about in the individual endothelium by means of mild localized damage to the endothelium with x-ray contrast media (Gospos et al. 1981) or following an excessive generalized injury of the endothelium produced by endotoxin shock (Freudenberg 1978). The degree of injury can only be determined by quantitative investigations which use standardized sampling procedures. The following changes are considered to be qualitative ultrastructural signs indicating a degeneration of the endothelial cell due to acute injuries: "Blebs" (Fig. 4), increased osmiophilia of the luminal plasmalemma (Fig. 5), vacuolar degeneration (Fig. 6), detachment of degenerated endothelium (Fig. 7) and intracytoplasmic oedema. In addition we were able to demonstrate morphometrically under the electron microscope an acute degenerative alteration of the endothelium. Fig. 8 shows in the upper portion (Fig. 8a) the intima of a central hepatic vein which reveals degenerative endothelial changes 6 hours after shock inducing endotoxin injection. The lower histogram (Fig. 8b) summarizes the results of ultrastructural morphometry from this experiment. The endothelium of the hepatic central vein shows the following alterations:

1. decrease of dense bodies (Weibel-Palade bodies included)
2. decrease of mitochondria
3. decrease of the rough endoplasmic reticulum and
4. increase of "autolytic" vacuoles

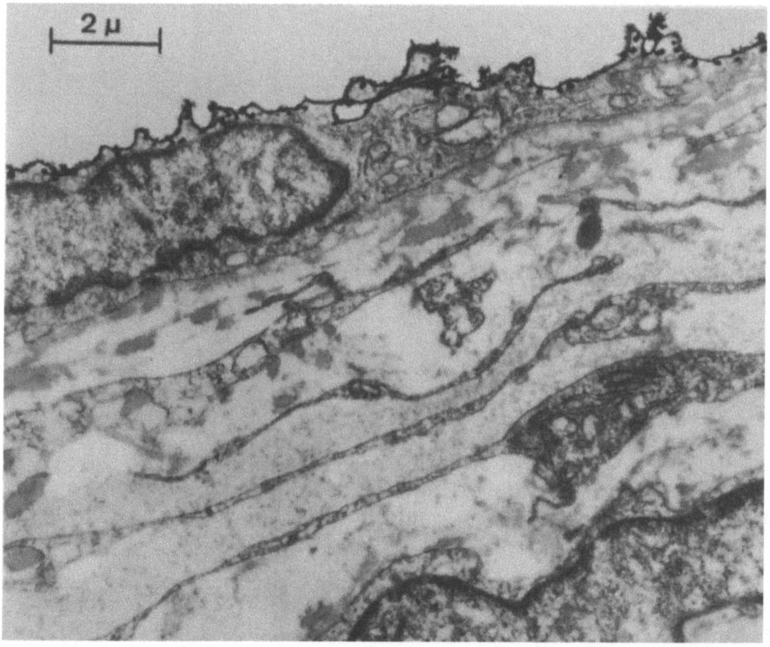

Fig. 5. Increased osmiophilia of the luminal plasmalemma and of the apical micropinocytotic vesicles of an endothelial cell, probably indicating a membranous injury of the endothelial cell. Rabbit mitral valve, 24 hours after induction of generalized Schwartzman reaction. TEM (from Freudenberg et al. 1982)

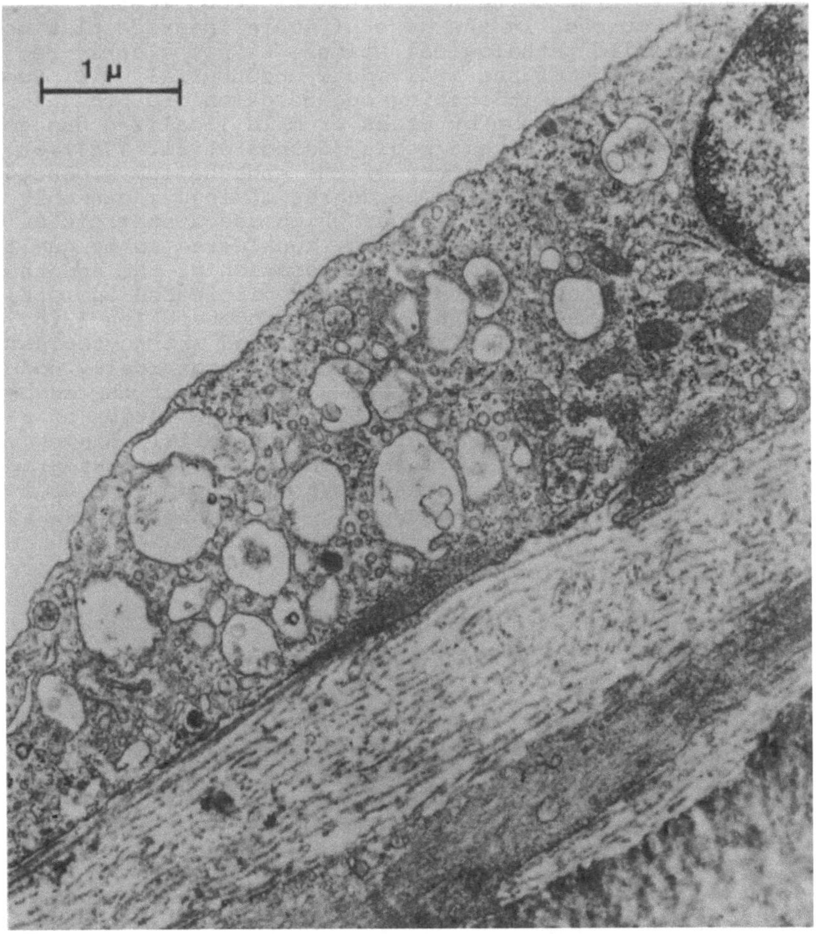

Fig. 6. Vacuolar degeneration of an endothelial cell indicating cellular injury. Renal artery of minipig. TEM (photograph lent by Prof. Staubesand)

Both acute and chronic injury may lead to osmioendothelial nécrosis. Changes of endothelial intercellular junctions and fatty degeneration are considered to be indications of chronic damage to endothelium.

Ultrastructural signs indicating an activation of endothelial cells were observed in regenerating endothelium following injury. These endothelial cells show a hyperplasia of their organelles, particularly an increased amount of rough endoplasmic reticulum and of Golgi-apparatus.

Classification of Endothelial Injury

The method used in our laboratory for classification of the extent of endothelial damage is based both on the morphological evidence of cellular injury and on the estimation of the proliferation process of en-

Table 2. Classification of the damage produced in the endothelium by endotoxic shock of degrees I - IV. Estimation of the proliferation quotient

$$P_Q = \left(\frac{\text{maximum } ^3\text{H-thymidine index after endotoxin administration}}{^3\text{H-thymidine index of controls}} \right)$$

and degree of damage compared with proliferation in controls (^{3}H-thymidine index and homogeneous or non-homogeneous proliferation pattern)

Degrees of damage	Vascular region	P_Q	Controls ^{3}H-thymidine index	Proliferation pattern
	Thoracic duct	7	$\rightarrow$	homogeneous
	Renal veins	8	$\rightarrow$	non-homogeneous
	Liver veins	8	$\rightarrow$	homogeneous
	Pulmonary arteries	9	$\leftarrow$	homogeneous
	Hepatic arteries	9	$\leftarrow$	homogeneous
	Renal arteries	10	$\leftarrow$	non-homogeneous
I	Small pulmonary vessels	10	$\leftarrow$	homogeneous
	Central hepatic veins	10	$\leftarrow$	homogeneous
	Inf. vena cava (prox.)	12	$\leftarrow$	non-homogeneous
	Pulmonary capillaries	12	$\uparrow$	homogeneous
	Pulmonary veins	13	$\leftarrow$	non-homogeneous
	Medullary capillaries of kidneys	14	$\rightarrow$	non-homogeneous
	Liver sinusoids	14	$\rightarrow$	non-homogeneous
II	Abdominal aorta	20	$\leftarrow$	non-homogeneous
	Abdominal aorta of juvenile rats (2 months old)	24	$\leftarrow$	non-homogeneous
III	Thoracic aorta of juvenile rats (2 months old)	25	$\leftarrow$	non-homogeneous
	Inf. vena cava (dist.)	27	$\downarrow$	homogeneous
IV	Thoracic aorta	33	$\leftarrow$	homogeneous

$\downarrow$ = slightly proliferating endothelium
$\leftarrow$ = moderately proliferating endothelium; ^{3}H-thymidine index < 0.495
$\rightarrow$ = moderately proliferating endothelium; ^{3}H-thymidine index > 0.495
$\uparrow$ = excessively proliferating endothelium

Homogeneous = homogeneously proliferating endothelium: the standard deviation is less than half of the mean value of the ^{3}H-thymidine index

Non-Homogeneous = non-homogeneously proliferating endothelium: the standard deviation is more than half of the mean value of the ^{3}H-thymidine index. From: Freudenberg et al. (1982)

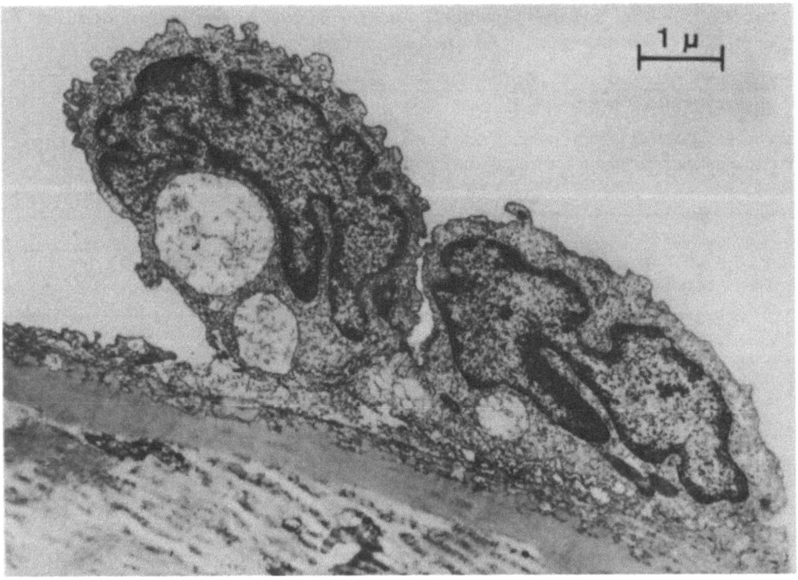

Fig. 7. Two degenerating endothelial cells showing cytoplasmic vacuoles near the nucleus and apical "blebs". The cell of the left side appears to be coming away from the intima. Rat aorta, 7 days after endotoxin administration. TEM (photograph lent by Prof. Staubesand

dothelium. Considering statistical differences in endothelial proliferation following intimal damage we classified the extent of injury to endothelium produced by an excessive pathological stimulus (endotoxin shock) by dividing it into 4 groups (Table 2):

- Group I: Slight damage
- Group II: Moderate damage
- Group III: Severe damage
- Group IV: Excessive damage

We found here that the former vascular endothelium which was proliferating slightly or non-homogeneously under physiological conditions tends to be more severely damaged than does the endothelium which is already proliferating intensely and/or homogeneously under physiological conditions.

Repair process

Our investigation inclines us to believe that the repair process of the normal endothelial tissue under physiological conditions (physiological repair) should be distinguished from the process of repair following pathological disturbances (pathological repair). On the whole, the same types of cells participate on both processes. A distinction between physiological and pathological repair of endothelium can be seen, amongst other things, in an intensified process of cellular regeneration (especially of proliferation) during pathological repair. We consider the increased appearance of white blood cells on the damaged endothelial surface as the earliest morphological sign indicating endothelial repair. Significant differences (depending on the vascular

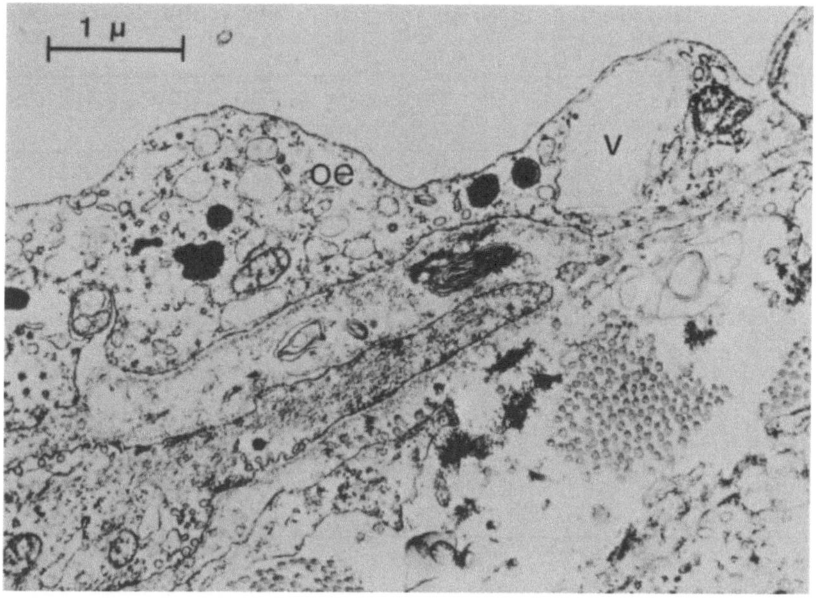

Fig. 8a. Intima of central hepatic vein 6 hours following shock-inducing endotoxin injection. Note the intracytoplasmic oedema (oe) and the increased number of autolytic vacuoles (V). TEM (from Freudenberg et al. 1982)

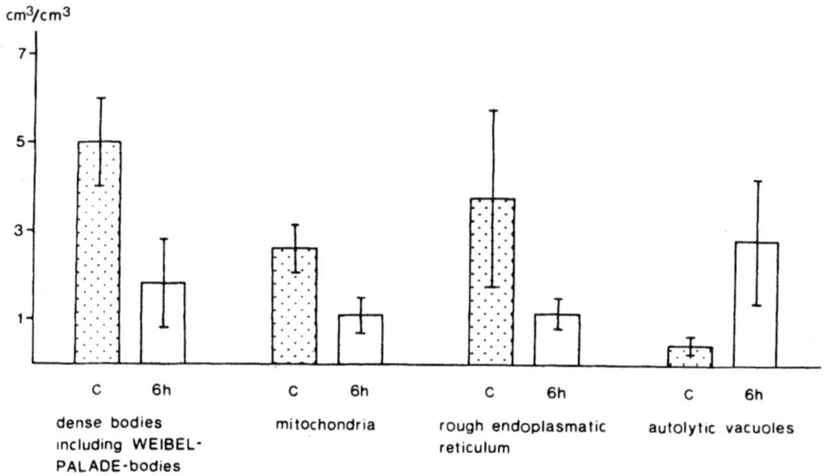

Fig. 8b. The volume densities (cm^3/cm^3; determined by ultrastructural morphometry) of "dense bodies" (including Weibel-Palade bodies), mitochondria, RER and "autolytic" vacuoles in the endothelium of the central hepatic veins of control animals (c) and of animals injected with endotoxin 6 hours before (6h) (from Freudenberg et al. 1982)

Table 3. Alterations of cell composition in the "Häutchen" preparation from different vascular regions following shock-inducing endotoxin administration (from Freudenberg et al. 1982)

	Controls		2nd - 5th day after endotoxin administration	
Vascular region	Endothelial cells	Non-endothelial cells	Endothelial cells	Non-endothelial cells
Aorta of juvenile rats	91 %	9 %	70 %	30 %
Aorta of adult rats	90 %	10 %	70 %	30 %
Inf. vena cava (dist.)	95 %	5 %	94 %	6 %
Inf. vena cava (prox.)	96 %	4 %	94 %	6 %
Thoracic duct	99 %	1 %	99 %	1 %

section) in the numerical participation of non-endothelial cells in this process may be observed in Table 3. We found here an increased number of non-endothelial cells, particularly markedly in the aortic intima, during the repair process of endothelium. During this period, both endothelial cells and mononuclear cells in the intima reveal signs of increased metabolic activity:

1. DNA synthesis of the intimal cells is enormously increased compared with the controls. We were able to show this increase by means of ^{3}H thymidine autoradiography in the "Häutchen"-preparation.
2. The enzyme activity of the naphtol-AS-acetate esterase in endothelial cells, monocytes and macrophages of the intima is also significantly increased during this period.

Conclusions

The rather confusing multiplicity of individual properties now known to be attributed to the endothelium allows two alternative conclusions to be drawn. Either the endothelial cell must be considered as one of the most pluripotent cells seen in the organism (Thilo and Heinrich 1981), or a great number of endothelial cells exists, having an enormous variety of individual functions. We consider ourselves in a position to assume that the endothelial cells represent a cell system which plays a highly significant role in the interaction of all systems of the organism.

References

Freudenberg N (1978) Endothelium and shock. Path Res Pract 162: 105-114

Freudenberg N, Riese KH, Freudenberg MA The vascular endothelial system. Structure - function - pathology - reaction to endotoxin shock - methods of investigation. Fischer, Stuttgart (in press)

Gospos CH, Freudenberg N, Staubesand J (1981) Wirkung von Kontrastmitteln auf das Aortenendothel. Pathologe 2: 247

Jaffe EA, Nachman RL, Becker CG, Minick CR (1973) Culture of human endothelial cells derived from umbilical veins. Identification by morphologic and immunologic criteria. J Clin Invest 52: 2745-2756

Riese KH, Freudenberg N, Haas W (1978) "En face" preparation methods for investigation endothelia and mesothelia. Path Res Pract 162: 327-336

Schwartz SM, Benditt EP (1973) Cell replication in the aortic endothelium: a new method for study of the problem. Lab Invest 28: 699-707

Sinapius D (1952) Über das Aortenendothel. Virchows Arch Path Anat 332: 662-694

Steinsiepe KF, Weibel ER (1970) Elektronenmikroskopische Untersuchungen an spezifischen Organellen von Endothelzellen des Frosches (Rana temporaria). Z Zellforsch 108: 105-126

Thilo DGS, Heinrich D (1981) Die Endothelzelle - eine pluripotente Kontrollzelle der Gefäßwand. In: Fischer H, Betz E (Hrsg) Das Gefäßendothel. Vorträge der Arbeitstagung des Tübinger Arbeitskreises für Gefäßerkrankungen e.V. Berchtesgaden, 13./14. Juni 1980. Wissenschaftliche Verlagsgesellschaft, Stuttgart

Discussion

BORN: What do you believe to be the most reliable quantitative indication of cell turnover when you look at large numbers of cells on the surface?

FREUDENBERG: We have shown this with a Häutchen method. By the Häutchen method you can get a monolayer of cells, and you have no intimal constituents, no subintimal constituents, only the endothelial cells and the other intimal cells. Therefore you can make a very correct counting of all cells of the intima.

SPRAGUE: You mentioned the lipoprotein receptors on the endothelial cells. I would just like to ask you how frequently do you actually see the so-called "coated pit" on intact endothelium? There are lipoprotein receptors associated with endothelium and usually that phenomenon is associated with the coated pit along the endothelial surface.

FREUDENBERG: We did not investigate this subject, but have only given here a literature review; based on these, we mainly worked with morphology and the growth pattern of endothelium.

C. SCHWARTZ: I guess what Dr. Sprague was really getting at is that in his experiments and ours there are few coated pits on the endothelium.

S. SCHWARTZ: I am not aware that anyone has a proportionality constant between the number of LDL receptors in cells and the number of coated pits. There is no particular reason why they should be proportional. The coated pit is important for almost all receptor mediated binding, not just for LDL. The endothelial cell in fact is not particularly rich in LDL receptors. What one sees a lot is the modified LDL receptor and the excitement about that now is Steinberg's observation that the endothelial cell, at least in human and baboon endothelium, is capable itself of modifying LDL so that it then binds to the modified LDL receptor. But there are coated pits in the endothelium. How many I do not know.

C. SCHWARTZ: I think there are not as many as one had imagined there would be. I think this is really the point.

SPRAGUE: Yes, we have looked at a lot of micrographs from the baboon aorta and found them to be extremely rare in that particular case.

BLACKSHEAR: Can you tell us anything about the stress of adhesion of the cell to the sub-endothelium in health and after injury or the distribution or change in adhesion throughout the vascular tree?

FREUDENBERG: I showed one picture where you can see the endothelium detached, but it is a very single observation and we only saw it some minutes after endotoxin application.

BLACKSHEAR: Would you think that, if you measured the force required to pull the Häutchen preparation off, this might be representative of the force that might be required.

FREUDENBERG: By the Häutchen preparation you cannot make this quantitation for this technique allows one only to say whether it is a good monolayer or not.

The Endothelium as a Mediating Factor in Transport Across the Arterial Wall

S. Weinbaum, G. Tzeghai, P. Ganatos, and R. Pfeffer

A. Introduction

Uptake studies with labeled tracer molecules have convincingly demonstrated that the endothelial cell layer is the principal resistance barrier in the transport of macromolecules across the artery wall [1,2]. In vivo experiments with pig aorta have shown that there are localized regions of enhanced macromolecule permeability [3]. These regions which stain more readily with protein binding Evans Blue dye than the unstained "white" areas have also been shown to exhibit a significant increase in cell turnover [4]. The authors have hypothesized that the localized increase in uptake observed in these in vivo experiments is due to one of the following mechanisms: (a) increase in permeability of the endothelial junctions, (b) changes in the vesicular transport rate and (c) increase in endothelial permeability due to enhanced cell turnover. The likelihood of each of these mechanisms will be examined with the aid of mathematical models and supporting laboratory experiments described by our collegue Dr. Shu Chien in a companion paper on this program.

Mechanisms (a) and (c) above are thought to be closely related. Electronmicroscope studies habe shown that the process of cell turnover associated with naturally occurring hemodynamic factors does not cause denudation [5]. This suggests that when a cell dies its junctional complexes become disrupted and leaky while the healthy cells around the dying cells slowly slough it off. While hemodynamic forces do not cause significant denudation of the endothelial cell layer at physiological shear rates it is possible that they can (i) effect the cell turnover rate and (ii) cause a migration of the integral protein strands that are the principal structural components of the cell junction complexes. Several investigators have demonstrated that the orientation of the cell and its junction strands is determined in large measure by the direction of fluid shearing stresses.

Experimental studies on the movement of water across the arterial wall due to a pressure driving force and the hydraulic conductivity of the various layers of the wall suggest that as much as half of the total hydraulic resistance occurs across the endothelial cell layer [6]. This finding is rather surprising considering that the thickness of the endothelial cell layer is less than 1/1000 the thickness of the thickness of the arterial wall media for the larger arteries of most mammals and humans. A mathematical model has been proposed to provide fundamental insight into the fine structure of the water movement and pressure distribution in the arterial wall, and in particular, to show quantitatively how the dimension of the intercellular cleft between adjacent endothelial cells is related to both the water movement and pressure distribution in the subendothelial space.

Fluid Dynamics as a Localizing Factor for Atherosclerosis
Editors: G. Schettler et al.

B. Vesicular Transport

The in situ experiments and vesicle transport theory reported in [7] and [8] provide important evidence that vesicle transport is not substantially increased by small amplitude mechanical oscillations. The theoretical model relates the vesicle attachment and transendothelial diffusion times to the number of free and attached vesicles. The theory and experiment support the hypothesis that vesicle attachment/detachment as opposed to vesicle diffusion is the rate limiting process and that this process is not significantly effected by mechanical disturbances. Elevated transmural pressure on the other hand was found to produce an increase in albumin uptake that did not vary linearly with increased luminal area [7]. A transmural pressure of 200 mmHg produced a fourfold increase in albumin uptake above zero pressure controls whereas the luminal area increased approximately twofold. Similar experimental findings are also reported in a recent paper by Leszczynski and Kummerow [9]. The theoretical model predicts that about 30 percent of the nonlinear increase is due to the narrowing of the endothelial cell layer at elevated pressures and the increased likelihood that a vesicle will cross the cell once it is released. The additional increase in uptake beyond this might be due to the widening of the intercellular cleft between adjoining endothelial cells at elevated pressures. The mathematical model described in the next section strongly supports this conjecture.

C. Endothelial Cell Turnover

The new model for endothelial cell turnover described in this section has provided a very plausible mechanism to explain the large local differences in albumin permeability that have been observed in vivo. In previous models of transport of macromolecules across the arterial wall [10,11] it was assumed that a damaged or dying endothelial cell offered no resistance to transport; thus the concentration underneath such a cell was assumed to be the same as in the lumen. The uptake of LDL and related macromolecules in the wall was explained by considering both the vesicular transport at the endothelial interface as well as diffusional transport in the underlying tissue. The recent experiment evidence [5] described in section A supports the view that dying endothelial cells do not leave demonstrable areas of exposed tissue but rather that the junctional complex around such cells may become leaky to macromolecules during the cell turnover process. As suggested above a similar process might well occur at elevated transmural pressures as the endothelium becomes sufficiently attenuated. Since under normal in vivo conditions 0.1 to 1.0 percent of the endothelial surface is involved in cell turnover and the intercellular cleft is only 0.1 percent of the cell surface area, the feasibility of the leaky junction hypothesis as a mechanism of enhanced transport has to be critically examined on quantitative grounds. The present model examines this potentially important mechanism by computing the increase in uptake and flux due to leaky junctions as a function of cell turnover.

The main difference between the new model described herein and the previous models [10,11] which study the effect of local endothelial damage is the assumption that when an endothelial cell is in the process of turnover, it becomes leaky only along its perimeter and macromolecules larger than 40 Å can be rapidly transported through the disrupted junctional complexes between dying and normal cells. This model like the previous ones is a mixed boundary value problem. However, since the width of a leaky junction is very small (10^{-5}) when normalized to the wall thickness, special solution techniques had to be developed to find the fine structure of the solution in the vicinity of the junctional complex.

A schematic diagram of the geometry of the problem is shown in Fig. 1, where R_1 is the radius of a damaged or dying endothelial cell, R_2-R_1 is the width of the leaky junction and all lengths are nondimensionalized by the thickness of the arterial wall. The arterial media is represented by a dispersed cellular phase and interstitial fluid phase as in the previous models. The solution for the steady-state concentration distribution as a function of r and z is obtained by representing the total concentration of the diffusing species as a superimposition of two contributions in which, C_1 corresponds to the solution for undamaged endothelium and C_2 accounts for the damage, and second by requiring that the sum of C_1 and C_2 satisfies the boundary conditions shown in Fig. 1.

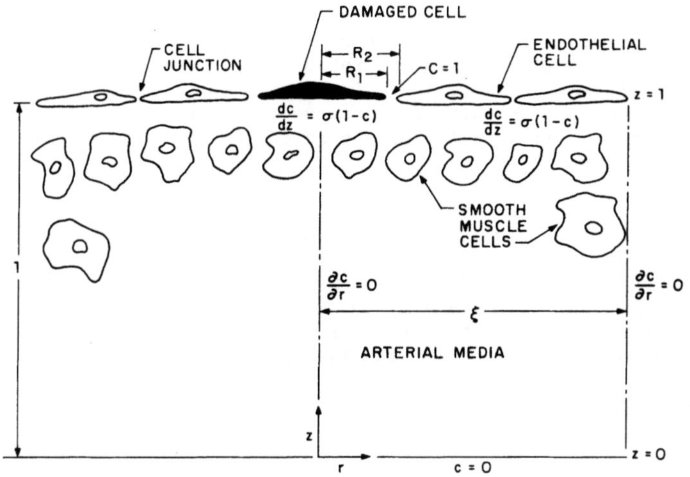

Fig. 1. Schematic of mathematical model of artery wall with endothelial cell turnover. Dying cells are periodically distributed and have leaky junctional complexes of radius R_2-R_1

Fig. 2 shows the principal results, the ratio of the steady state flux ψ/ψ_i for a damaged to an intact endothelial cell layer, as a function of the fraction of damaged cells or cell turnover rate ϕ for a leaky junction whose width is 200 Å, a Biot number σ of 0.2 and a wall thickness of 1 mm. Here the Biot number represents the ratio of the resistance to transport of macromolecules offered by the arterial media to the resistance of the intact endothelium. Figure 2 also predicts the ratio of the uptake in the artery wall with a damaged and intact endothelium since in the steady state the uptake and flux ratios are equal to one another. It is interesting to note as the number of damaged cells or turnover rate increases from 1/1000 to 5/1000, the flux or uptake ratio increases roughly by 67 percent clearly indicating the rate controlling effect of the endothelial cell monolayer. Thus large changes in the equilibrium balance of the cholesterol carrying LDL in the arterial wall can occur at very small cell turnover rates.

Also shown in Fig. 2 (dashed curve) are the results for the previous model reported in [11] in which the dying endothelial cell was removed and treated as an open hole. The closeness of the results for the 200 Å cleft and the model for the hole show that the transport rate for an artery wall of 1 mm thickness and specified cell turnover rate

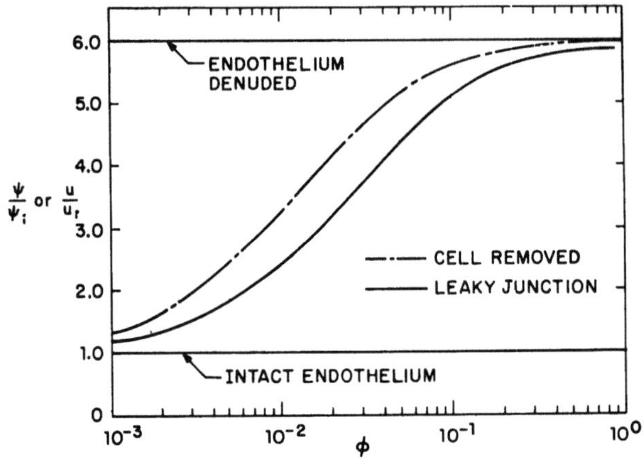

Fig. 2. Ratio of flux ψ or uptake in an artery wall with endothelial cell turnover to an artery with intact endothelium (subscript i) where transport occurs only through transendothelial vesicle migration. ϕ is the ratio of dying to healthy endothelial cells. Solid curve theoretical results for a leaky cell junction around dying cell of 200 Å width. Broken curve results for same geometry but dying endothelial cell completely removed leaving circular hole exposed to concentration in lumen. R, 15 µm, wall thickness 1 mm

is largely controlled by the perimeter of the dying cell rather than its area. The solution for an intercellular cleft of 100 Å thickness is nearly indistinguishable from the 200 Å solution shown. These results show that the transport is relatively insensitive to the width of the intercellular cleft provided its dimensions are sufficiently large to permit the rapid passage of the diffusing molecule.

Bell et al. [3,4], reported experimental findings of a 66 percent increase in albumin influx rate $(\psi/\psi_i)\phi = 0.3/(\psi/\psi_i)\phi = 0.007)$ in the blue stained area of a pig aorta as compared to the unstained or white area. If a Biot number of 0.2 is assumed, using an endothelial cell length of 30 µm [4], and artery wall thickness of 1.0 mm and referring to Fig. 2, it is found that $(\psi/\psi_i)0.03/(\psi/\psi_i)0.007 = 1.76$ or a 76 percent increase. Thus it appears that the leaky junction hypothesis is not only a qualitatively reasonable explanation of the increased uptake found in certain regions of the arterial tree, but also is in favorable quantitative agreement with experimental physiological results.

D. Water Movement

Fig. 3 is a schematic diagram showing our mathematical model (not to scale) for pressure filtration across the artery wall. The important unknown about which there has been much discussion is the steady state pressure distribution in the tissue and, in particular, in the subendothelial space P_2. In modeling this problem, we first solve a porous media diffusion equation for the pressure field, and then use Darcy's Law to obtain an expression for the volumetric flow across the arterial media in terms of the unknown pressure P_2 just beneath the endothelium. A second equation is found by applying a lubrication type analysis to the fluid motion within the intercellular clefts and relating the volumetric flow rate to the pressure drop along the clefts. These equations are solved simultaneously to yield the flow across the arterial wall and the pressure distribution beneath the endothelium and

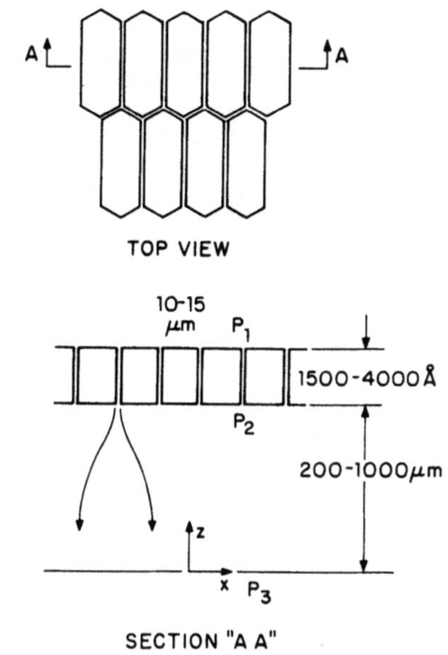

Fig. 3. Schematic diagram for mathematical model for water filtration across arterial endothelium and media

in the arterial media. The intercellular space is treated as a parallel channel with a localized constriction of length w which narrows from 200 Å to $2\varepsilon_{min}$ in the region of the junctional complex as shown in the inset of Fig. 4. The lubrication theory analysis is then used to express the hydraulic resistance of the intercellular cleft in terms of an equivalent parallel wall channel of width $2\varepsilon_0$. The right hand axis shows the relationship between ε_0 and ε_{min} for a sinusoidal constriction of 1000 Å length.

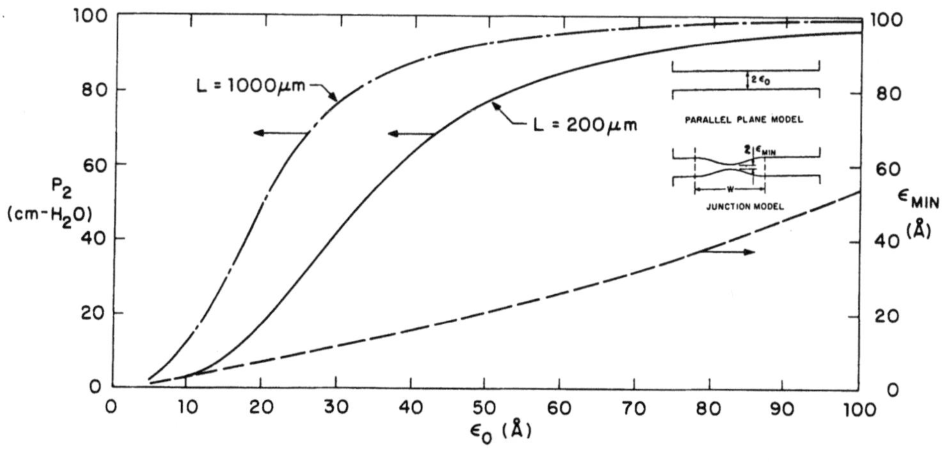

Fig. 4. Subendothelial pressure as a function of effective intercellular cleft width for 0.2 and 1.0 mm thick arterial walls. $2\varepsilon_{min}$ is the minimum width of junctional complex. ε_0 equivalent width of parallel walled channel having same hydraulic resistance as junction model shown in inset. Results shown based on intercellular cleft of 1 μm length, w = 1000 Å and sinusoidal constriction

The left hand axis of Fig. 4 shows the solutions for the pressure at the entrance to the subendothelial space for a transmural pressure of 100 cm water. One observed that when the effective intercellular cleft width $2\varepsilon_0$ is roughly 35 Å, more than one half of the total transmural pressure drop can be sustained by an endothelial cell layer for a vessel whose wall thickness is less than 1 mm. This effective width of the intercellular cleft is typical of the largest molecules that have been observed to traverse the endothelium via this pathway under normal in vivo conditions. For a thinner wall such as a rabbit aorta (L = 200 µm), the model predicts that roughly 80 percent of the hydraulic resistance resides in the endothelium. The results in Fig. 4 are based on the measured hydraulic conductivity of a denuded artery wall given in Vargas et al. [6].

A fascinating prediction of the model is the detailed solution for the water movement in the subendothelial space. Representative results for the velocity prolifes are shown in Fig. 5 for a wall of 200 µm thickness. The water upon exiting from the intercellular clefts quickly diffuses in planes parallel to the surface. The exit mixing between jets is therefore extremely rapid and a nearly uniform one-dimensional velocity profile is achieved after the water has penetrated only a few percent of the thickness of the arterial media. Since the local resistance of a porous medium is according to Darcy's law proportional to the local flow velocity much larger resistances are encountered in the thin region of rapid velocity variation than in the deeper regions of the media where the flow is one-dimensional. The filtration resistance of artery wall with intact endothelium therefore does not scale linearly with wall thickness. The behavior is confirmed by the experimental measurements [6]. A more detailed account of the theory and a parametric analysis of the relative hydraulic resistance of the endothelium and arterial media as a function of cleft geometry and wall thickness is given in [12].

This research was supported by NSF Grant ENG 78-22101 and NHLBI Grant HL 19454.

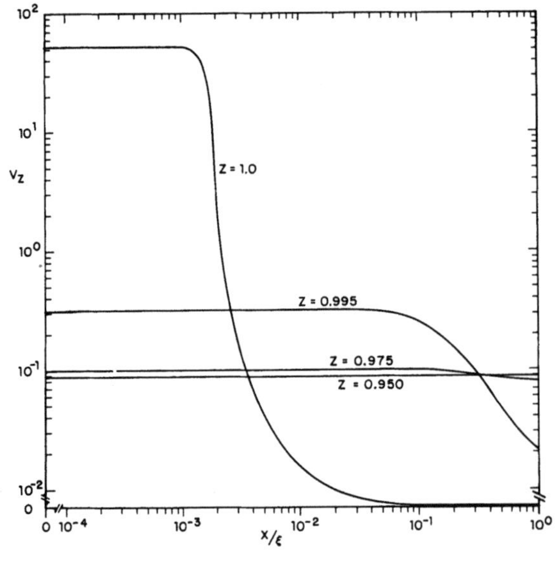

Fig. 5. Velocity profiles in horizontal planes as a function of dimensionless depth in the artery wall. x/ξ dimensionless horizontal coordinate. $x/\xi = 10^{-3}$ edge of intercellular cleft; $x/\xi = 1$ centerline of endothelial cell. Results shown are for wall thickness of 200 µm

References

1. Fry DL (1972) Localizing factors in arteriosclerosis. In: Likoff W, Segal B, Insull W, Mayer JH (eds) Atherosclerosis and coronary heart disease. Grune and Stratton, New York, pp 40-83
2. Siflinger A, Parker K, Caro CG (1975) Uptake of ^{135}I-albumin by the endothelial surface of the isolated dog common carotid artery. Cardiovasc Res 9: 478-489
3. Bell FP, Adamson JL, Schwartz CJ (1974) Aortic endothelial permeability to albumin: Focal and regional patterns of uptake and transmural distribution of ^{131}I-albumin. Exp Mol Path 20: 281
4. Caplan BA, Schwartz CJ (1973) Increased endothelial cell turnover in areas of in vivo Evans blue uptake in the pig aorta. Atherosclerosis 17: 401
5. Schwartz SM (1980) Role of endothelial integrity in atherosclerosis. In: Nerem R, Guyton J (eds) Hemodynamics of the artery wall. Proceedings NSF Specialists Mtg., University of Houston, Houston, Texas, Nov. 5-7
6. Vargas CB, Vargas FF, Pribyl JG, Blackshear PL (1978) Hydraulic conductivity of the endothelial and outer layers of the rabbit aorta. Am J Physiol 236: H53-H60
7. Chien S, Weinbaum S (1980) Vesicle transport in arterial endothelium and the influence of mechanical factors on macromolecular permeability. ASME J Biomech Eng 103: 673-684
8. Chien S, Lee ML, Laufer S, Handley DA, Weinbaum S, Caro CG, Usami S (1981) Effects of oscillatory mechanical disturbance in macromolecular uptake by arterial wall. Atherosclerosis 1: 326-336
9. Leszczynski DE, Kummerow FA (1981) Effects of normal stress on aorta surface stretching permeability. J Biomech 14: 307-314
10. Nir A, Pfeffer R (1979) Transport of macromolecules across arterial wall in the presence of local endothelial injury. J Theor Biol 81: 685-704
11. Pfeffer R, Ganatos P, Nir A, Weinbaum S (1981) Diffusion of macromolecules across the artery wall in the presence of multiple endothelial injuries. ASME J Biochemical Eng 103: 197-203
12. Tzeghai G, Ganatos P, Weinbaum S, Pfeffer R Theoretical model for steady state water movement and pressure distribution in the arterial wall (in preparation)

Discussion

S. SCHWARTZ: You were looking at two kinds of data - thiamine index data and plaque cells with silver stain. I will show you later that there is actually quite good evidence now that the frequency of truly dead cells in the endothelium is about 10 times the frequency of replicating cells seen at one time. I would think that the frequency of dead cells is the number that you would want.

WEINBAUM: In the normal white area what would you estimate to be the number of cells which would have leaky junctions or be dying?

S. SCHWARTZ: I do not do white versus blue, but in our data the frequency of cells that I would think have that property would be about ten times the thiamine index.

WEINBAUM: So it would be about one cell in a hundred then?

S. SCHWARTZ: If the replication frequency in that area is 0.1.

FRIEDMAN: I think there has been quite a bit of work done on the solution of the problem of a particle fairly close to the size of a pore through which it is passing. Also, I do have a question about the model. What I think you are telling me based on the model is that, once you have a sufficient amount of perimeter, you are going to get the same answer irrespective of whether you have more perimeter and irrespective of the configuration of that perimeter. If that inference is correct, then what the model is saying is that one cannot use these kinds of experiments to differentiate among the various hypotheses as long as those previous two conditions are met. Finally, have you considered or have you done any calculations in which you have included an anisotropic sub-endothelial space? We know that this region is organized in such a fashion that one might expect the diffusion constants to be different in the Z direction than in the R direction.

WEINBAUM: With regard to the first question, the solution of the particle in a pore problem had only been worked out for particles that are almost the same size as the pore or for a particle moving along a center line. The solution which I presented here is one people have been trying to do for 50 years. As for the second question, the interpretation that you have is both right and wrong as far as perimeter limited diffusion is concerned. The behavior which you saw here occurs only in the limit where the wall is thick compared to the radius of the circle. If it becomes smaller, what happens is that the details of the damage become important relative to the dimensions of the wall. The insensitivity came about from the fact that the wall is thick compared to the perimeter. If I made the perimeter bigger, then the permeability would have changed very substantially. The model is highly selective in the sense that it will tell you exactly how much turnover is required to give you a certain amount of increase in uptake. As long as we have a reference value for a white area, it will tell you for any other part of the vascular system what the turnover would have to be in that part of the system to give you the increased permeability that you would measure in a macroscopic experiment. Finally, an anisotropic subendothelium has been put into our models and also variable diffusion coefficients in the wall. The numbers are essentially the same. It is a minor variation; it does not change the mechanism. This was based on models that we have for the wall. In terms of the variability in the structure of the wall, we only have the results which I showed you yesterday. Those are the first concentration measurements to show you the non-isotropic nature of the diffusion. I do not believe there were any done before, and we are working on that now.

CORNHILL: I am a bit confused when we go from the Stokes jet through the phase between the cells to plug flow. Could you explain this a little more?

WEINBAUM: You are talking about a filtration flow. When you have a fluid coming out of a channel which is 150 angstroms wide, it covers a very, very tiny part of the sub-endothelial space. It has to become a one dimensional plug flow in the far field when the wall is thick. Now what happens here is that the cells are placed 10 microns apart, i.e. the dimensions of an endothelial cell. It could be 20 microns, but that dimension is small compared to the thickness of the wall, which is probably a 100 cells thick and so the fine structure of the periodic spacing of this jet would get wiped out over a length scale which is typical of the cell dimensions. When you do this problem, you first have a sub-structure where you have very complicated behavior occurring on a length scale of the order of the dimensions of the cell. Then you have another problem which is basically a one-dimensional problem when you get outside of that region. The reason that you could not treat the problem properly is because a lot of the dissipation is occurring in the region where the fluid is moving fast. You have a viscous fluid moving through a high resistance layer, and the resistance is greater because the fluid velocity is much larger in the sub-endothelial space than when you penetrate further into the media.

CARO: Have you taken into account the fluid-solid interactions in your model? I think we all agree that the transport is through compressible, deformable material. In this case I think D will depend upon the interaction between fluid and solid, and also, it does not seem to me a necessary condition that there is symmetry either.

WEINBAUM: There is a transient phenomenon, and there the fluid space is changing as a function of time until it achieves some steady state. Once you have achieved the steady state, then you have basically a fixed geometry that you are working with, and you do not have to worry about the interaction. You do have problems in the interaction in the cleft, where the cleft dimensions can change as a function of the pressure in the cleft, and that is not in the model. However, eventually you achieve a fixed configuration. In that fixed configuration the elasticity is not changing any more, and there is just going to be a fixed resistance. The resistance changes in the time-dependent problem, not in the steady state.

Morphology and Fluid-Dynamics of Endothelial Cells at the Site of Vascular Bifurcation

H. Buss

Former studies by light microscopists with Häutchen preparations or with silver-impregnations indicate a relationship between the orientation of endothelial cells and local blood flow conditions. We wanted to confirm and extend these observations by systematic studies of the vascular system of rabbits with the scanning electron microscope. Recent work of other authors have revealed comparable results. Concerning the present results, we concluded that the shape and orientation of a given endothelial cell is the result of the shear forces of blood flow. Other morphologic parameters like thickness of the cells and their nuclei, shape of nuclei, cell borders, microvilli, etc. change locally. Unfortunately, we can not correlate exactly these various cell alterations with the events of blood flow.

We first examined endothelial cells from the inferior caval vein within the liver region, but distant from a mouth of a liver vein. Here the cells are characterized by straight cell borders and some short microvilli at the surface. The average length is 81 µ and the average width is 14 µ. This means, that these venous cells can be up to 100% longer and wider than those of the aorta. However, the closer one comes to the right atrium of the heart, the shorter the cells become until their length measures about 38 µ and their width 20 µ. We feel that this might be due to counterforces induced by the heart section.

At the entry point of the hepatic veins, an irregular pattern of short rounded, trapezoid or polygonal cells is observed. Their mean length is 27 µ and the mean width is 18 µ. This irregular pattern begins just inside the hepatic veins and comprises a small circumscribed area within and somewhat downstream from the entry point into the caval vein. A comparable pattern can be observed at the mouth of the renal veins and at the entry of the saggital sinus into the transversal sinus of the brains. This contrasts with the bifurcation of the inferior caval vein, i.e. the confluence of the iliac veins, where the endothelial cells of the two veins come together without any disturbance in their pattern. This might be a consequence of the narrow angle between the iliac veins whereas the blood streams of the other mentioned veins come together at almost 90°. If one assumes that cell shape is a resultant of shear forces, the well circumscribed area of polygonal cells should be that point where shear forces of the two directions are equal. Similar cell patterns at arterial bifurcations need, of course, another interpretation.

The situation at arterial bifurcations is quite different depending on the width of the branches. Fig. 1 shows two arterioles of about 30 µ width. The first is branching from the cerebral basilar artery. The pattern of endothelial cells remains undisturbed in the branching area. At the lateral angle the cells run directly into the branch, whereas they turn around the edge of the bifurcation. The second arteriole is a branch of the renal artery. Again, the mouth seems to be punched out without any disturbance of the intimal pattern within the

Fluid Dynamics as a Localizing Factor for Atherosclerosis
Editors: G. Schettler et al.

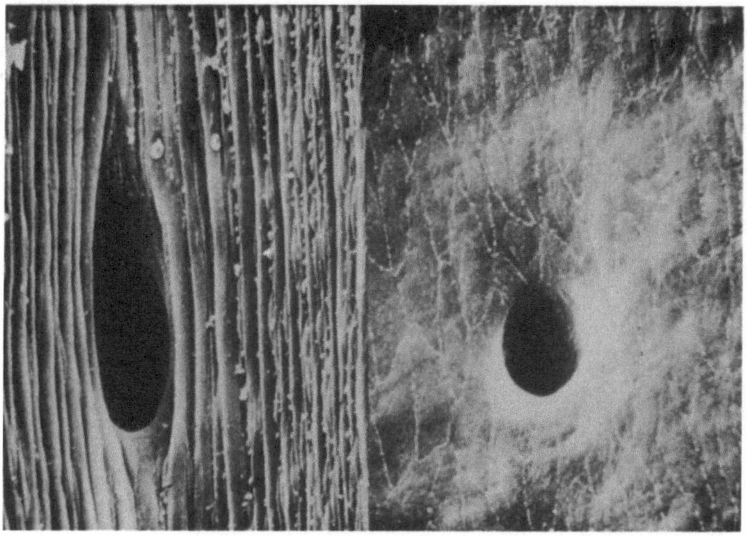

Fig. 1. SEMs from typical arterial bifurcations as taken at branches of the cerebral basilar artery and the renal artery

main artery. The endothelium is not as well demonstrated as in the cerebral artery because the renal artery has a prominent longitudinal relief which originates from the undulating internal elastic membrane and the state of contraction of the media muscle cells. However, the undisturbed relief reflects an unchanged inner structure of the arterial wall around the branching.

At the branching point of main arteries, the longitudinal relief divides into two parts, each running into one of the arteries. Proximal to the bifurcation between the dividing parts of the longitudinal relief, a triangle without such a relief is observed regularly. This is the site of the well known cushions with intimal fibrosis and splitting of the internal elastic membrane, and it is in this area where always a small group of polygonal or roundish cells can be detected as will be shown later. The divider is the usual form for greater arteries, that of a "U" or a horse-shoe. Distal to the rim of the flow divider, the longitudinal relief is somehow distorted and the cells seem to be swollen.

In the case of smaller arteries or in arteries without a prominent longitudinal relief, distal to the flow divider a trigon can be seen made up of short, polygonal cells. Cells of the usual shape are placed on both sides coming together at the distal point of the trigon. Fig. 2 derives from the coronary artery, and it represents another type of a flow divider which has the form of a six or of a spiral. This type is typical for medium and smaller branches of the coronary artery. Just behind the bifurcation, the artery is branching once more, and therefore the endothelial cells in depth run oblique to the axis of the first branch.

Fig. 3 demonstrates the corresponding area at the bifurcation of the renal artery. Here, the rounded cells make up a large area, they seem to be slightly swollen, and they possess protruding the nuclei.

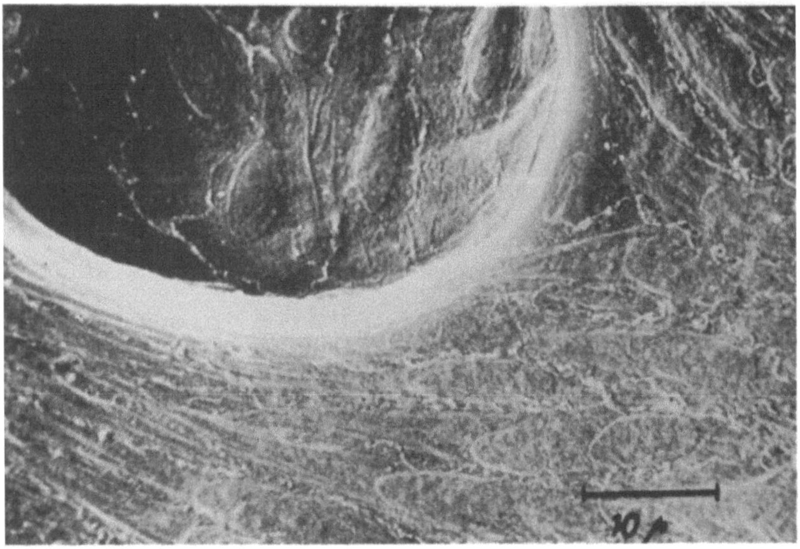

Fig. 2. SEM example from a branching of the coronary artery

Fig. 3. SEM photograph taken from a region at the bifurcation of the renal artery

With the help of serial pictures, we have tried to make schematic drawings of the main orientation of endothelial cells at some bifurcations. Fig. 4 demonstrates our findings around the bifurcations of the renal arteries. The cell orientation corresponds to the known theoretical figures of flow; i.e. the deviation of the main aortic flow is towards the branches, the dividing flow at the bifurcation and the secondary spiral flow within the first part of the branches. The small rings indicate the area of polygonal cells at the beginning of the divider.

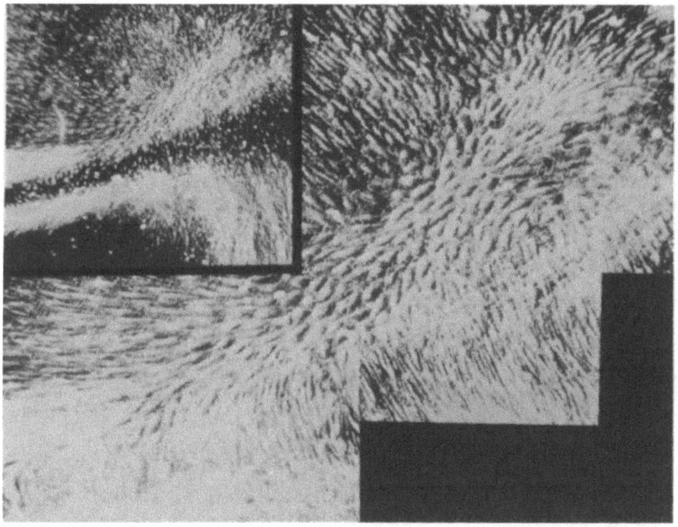

Fig. 4. Orientation of endothelial cells around the bifurcation of the renal artery

The deviation of cells from the axis of the renal artery due to secondary flow depends on the bifurcation angle. If the angle is about 90° as, for example, in the coeliac trunk, cells run with their long axis at almost a right angle to the lateral wall and come together with those cells oriented along with the primary flow of the lateral angle. Another area of polygonal cells or spindle shaped cells oriented in various directions may be observed at this site.

Analogous to this are the findings at the aortic arch. The secondary flow as indicated by cell direction goes down to the inner convexity. Here again you will find an area of polygonal cells. At the bifurcation of the carotid arteries, we did not find greater areas of polygonal cells. This might be a consequence of the very narrow angle of these branches.

At this point I would like to raise the question if the areas of polygonal endothelial cells are the morphologic substrate of only one flow condition? My answer is "No". We have found these cells at sites of confluent venous flows, at the raising point of the arterial flow divider, just downstream of the run of the bifurcation, and in some cases at lateral angles, possibly in regions of flow separation.

Another very large area of polygonal cells is the entire ascending aorta, an area where flow disturbance is a physiological event. These cells are more cuboid with protruding nuclei and form a cobblestone surface. They are obviously not identical with the flat cells within the more distal coronary artery branching. They represent the Evans blue positive area named by C. Schwartz "pre-lesion area". However, I do not believe that this is "pre-lesion"; it is rather physiologic and adaptive to the various flow forces acting together in this region.

The purpose of our work was to clarify the details of endothelial surface structure. However, several questions have been left unanswered. I feel that in recent time almost too much work has been done with models. Models can give only a vague impression of what is really happening along a branching artery with pulsating flow, changing elasticity and muscle tone, and with always changing anatomical sizes during action. The present work of morphologists concerning the interrelationship between blood flow, the endothelium, and the underlying structures can only be the beginning of broader research.

Discussion

STEHBENS: I appreciate that it is difficult to discern giant cells and multi-nucleated giant cells in endothelium using scanning electromicroscopy, but were you able to see any in your preparation?

BUSS: No. I do not see this. We have made very extensive observations on the venous system of rabbits and of the arterial system, and we never saw these cells.

STEHBENS: They are related to age. Did you use only young animals or did you else use older animals?

BUSS: No, these are adult animals of about 7 month to one year.

DEWEY: I am going to ask you a very unfair question. What happens to the venous endothelium in a coronary artery by-pass?

BUSS: We have not studied this, but if you mean the region where the flow is impinging upon the venous wall, there is loss of endothelium.

BLACKSHEAR: About the size of the venous endothelial cells, if you look at the veins, do you think you would be able to deduce the average venous shear stress by the size of the cells?

BUSS: Yes. That might be correct.

Location and Morphology of Hypertensive Lesions in Coronary Arteries of Dogs

S. Meairs, E. Weihe, U. Mittmann, and W. G. Forssmann

Summary

Hypertension was produced in coronary arteries of 54 foxhounds by afterload stress with supravalvular aortic stenosis. Functional hemodynamic studies were carried out after 3 months and one year of hypertension. Other dogs were perfusion fixed after 3 months and one year of hypertension for morphologic analysis of coronary arteries with scanning electron microscopy (SEM), transmission electron microscopy (TEM), and light microscopy (LM). SEM results of coronary arteries subjected to hypertension show little change from normal coronary arteries after three months. Only isolated cases of bright spindle-shaped cells which appear to be sloughing from the surface were seen directly proximal to ostia of coronary arteries. TEM showed no ultrastructural changes in this banding group. SEM observations of coronary arteries after one year of hypertension revealed areas of endothelial denudation oriented parallel to the longitudinal axis of the artery, a few cell diameters wide and 200-300 μ in length. These lesions were often seen to form narrow channels between two or more branching points, either in a continuous or in series fashion. Denudations were restricted to arteries greater than 1 mm in diameter. Platelets, lymphocytes, granulocytes, and monocytes adhered to the subendothelial surface. Also in proximity to branching points were heterogeneous populations of cells adherent to intact endothelium. TEM and LM of this group showed breaks and duplication in the intima. Focal areas of inflammatory cells and plasma insudation in the intima were noted in areas exhibiting endothelial surface adherence.

Introduction

Epidemiologic studies have consistently shown that hypertension is a major risk factor for the development of atherosclerosis [11,3]. Clinical trials have demonstrated that therapeutic reduction of blood pressure lowers the risk for atherosclerosis [6]. The actual manner in which hypertension contributes to atherogenesis has remained largely a matter of speculation and firm morphologic confirmation for hemodynamic theories is lacking. Several approaches have been utilized to investigate this problem. Hypertension has been produced with renal artery stenosis or unilateral nephrectomy with a high salt diet [12, 13]. These experimental models are unfortunately burdened by a complex interpretation of inadequately defined renal mechanisms which may obscure the effect of hypertension per se on lesion development. Indeed, results from such experiments indicate that endothelium may react differently to different types of hypertensive stimuli [12,13]. Moreover, the extrapolation of information obtained from studies of spontaneously hypertensive rats does not provide us with an answer to the question of the singular role of hypertension as a hemodynamic factor in atherogenesis. It is for this reason that we have developed a model to study the effect of hypertension on the coronary vascular system of foxhounds which allows a correlation of morphologic results with hemodynamic parameters. To produce a sustained and gradual development of hypertension in coronary arteries, foxhounds were subjected to supravalvular aortic stenosis. After 3 months and 1 year of hyper-

Fluid Dynamics as a Localizing Factor for Atherosclerosis
Editors: G. Schettler et al.

tension, the coronary arteries were studied with scanning electron microscopy (SEM), transmission electron microscopy (TEM), and light microscopy (LM). Functional hemodynamic studies were carried out and correlated with morphologic results.

Material and Methods

Banding Operation

The banding operation was performed on 54 foxhounds, 6-8 weeks old and weighing 3.0-5.5 kg. Premedication involved the use of N-dimethyl-aminopropyl-3-propionylphenothiazine (Combelen[R]), pentobarbital-sodium (Nembutal[R]), diallyl-bisnortoxiferine (Alloferin[R]), and morphine followed by nitrous oxide inhalation anesthesia. A thoracotomy was performed and a cervix ligature band (Ethicon[R]) was placed on the ascending aorta proximal to the brachiocephalic artery, gently constricted to obtain an initial systolic gradient of 10 mmHg, and then sutured. The thorax was closed and 1 million units of penicillin were given for infection prophylaxis.

Functional Hemodynamic Studies

Three months after the banding operation a catheterization was performed to determine prestenotic aortic pressure and pressure gradient across the stenosis. Detailed studies of myocardial blood flow, cardiac output, contractility, and coronary reserve were carried out on 13 dogs after 3 months of stenosis and on 9 dogs after one year of stenosis. These experiments are beyond the scope of this presentation and are discussed in depth elsewhere [23].

Perfusion Fixation

Four groups of six dogs each (control 3 months and 1 year; hypertension 3 months and 1 year) were selected for perfusion fixation to study the surface structure and ultrastructure of the coronary artery system. Premedication and anesthesia were the same as for the aortic banding operation. The thorax was opened and a cannula with a balloon attachment corresponding to the diameter of the aorta was inserted into the descending aorta for the retrograde perfusion of the heart. The heart was arrested in diastole by injecting a bolus of cardioplegic solution (Cardioplegin[R]) into the aorta and left auricle. The great vessels of the aortic arch were clamped to ensure optimal perfusion of the heart. Outflow was maintained by opening the inferior and superior vena cava and pulmonary veins and arteries. The coronary vascular system was rinsed at 4°C with a 300 mosmol solution (Kardioplegische Lösung nach Bretschneider[R]) containing procaine and heparin for 3-5 min at room temperature. Flow was maintained at 100 ml/100 g/min and perfusion pressure was adjusted to 60 mmHg in the ascending aorta. As fixative a 1.5% glutaraldehyde - 1.5% formaldehyde mixture in 0.1 M phosphate buffer (pH 7.35) with 25 g/l polyvinylpyrrolidone was perfused for 10 min [15].

Tissue Preparation

Coronary arteries were dissected and cut into small segments of 5 mm in length and systematically analyzed with SEM. Location and inner arterial diameter were noted and specimens for TEM and LM were taken from the same regions for comparison. For SEM, arterial segments were washed in 0.1 M cacodylate buffer, postfixed in 1% osmium tetroxide, buffered with 0.1 N sodium cacodylate pH 7.2, and dehydrated in alcohol series. Specimens were critical point dried with CO_2 directly from absolute alcohol. Specimens were mounted, coated with gold, and observed with the Philips PSEM 500. In selected cases SEM specimens were reembedded for TEM analysis to elucidate the ultrastructure of SEM findings.

For thin sections specimens were rinsed in maleate buffer, postfixed in cacodylate-buffered osmium, en bloc stained with maleate-buffered uranyl-acetate, dehydrated in alcohol and embedded in Epon-Araldite. Thin sections were cut on LKB-ultramicrotomes, stained with uranyl-acetate in acetone-aqueous solution and lead citrate, and observed with a Zeiss EM 10 electron microscope.

Results

Functional Hemodynamic Studies

Measurements of hemodynamic parameters after 3 months of stenosis showed a mean systolic pressure gradient of 31 ± 18 mmHg across the stenosis. The mean prestenotic pressures in mmHg were systolic 134 ± 12, diastolic 83 ± 8, and arterial 100 ± 11 as compared to normal aortic pressures of systolic 105 ± 12, diastolic 82.5 ± 9, and arterial 90 ± 12. The relationship between the left ventricular muscle mass and end-diastolic volume was increased and indicated left ventricular hypertrophy. Contractility was normal at rest and only slightly decreased with maximal pressure stress. Myocardial blood flow was normal, but coronary reserve was decreased. After 1 year of aortic stenosis (373 ± 12 days) changes in hemodynamic parameters were more dramatic. The mean systolic pressure gradient was 62 ± 45 mmHg. The mean prestenotic pressures in mmHg were systolic 185 ± 38, diastolic 104 ± 13, and arterial 136 ± 21. The cardiac output was reduced by 17% and the contractility was decreased by 26%. The resting myocardial blood flow was normal. The coronary reserve was significantly decreased. The mean prestenotic pressures for dogs perfusion fixed after 1 year of banding (362 ± 1 days) in mmHg were systolic 182 ± 31, arterial 128 ± 22, and diastolic 101 ± 16.

Morphology

Normal coronary artery endothelium is characterized by a uniform, rather smooth surface structure, exhibiting rare microvillous projections. Nuclear bulging is elongated in the direction of the longitudinal axis of the coronary artery. Only in the region of the ostium of the main branches ramifications are irregularities of cell shape noted. No differences exist between the left and right coronary arteries. As the microvasculature is approached the surface structure becomes smoother and more homogeneous. Coronary veins show uniform rhomboid cell shapes with distinct borders.

SEM results of coronary arteries subjected to hypertension show little change after three months of stenosis. Only isolated cases of bright spindle-shaped cells which appear to be sloughing from the surface were seen directly proximal to ostia of coronary arteries. Such cells were not seen in normal control dogs. TEM of this banding group showed no ultrastructural changes of coronary arteries.

SEM observations of coronary arteries from dogs subjected to one year of aortic stenosis revealed significant pathologic alterations. In each animal investigated several areas of endothelial denudation were discovered in proximity to branch points. These lesions were typically oriented parallel to the longitudinal axis of the artery, a few cell diameters wide and 200 - 300 μ in length (Fig. 1). These denudations were often seen to form narrow channels between two or more branching

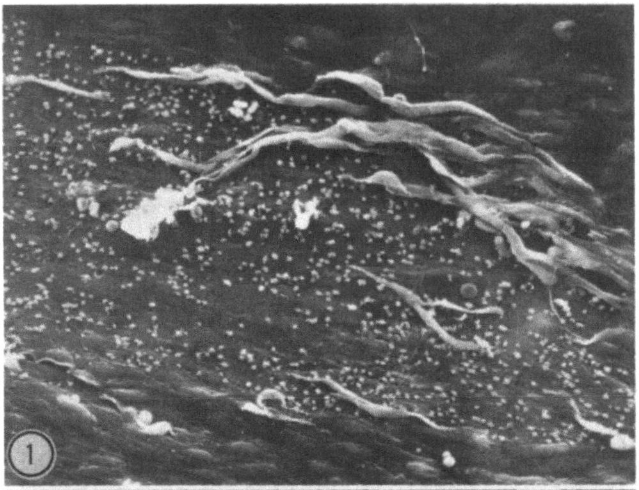

Fig. 1. Endothelial denudation. Platelets and blood cells adhere to subendothelial space. Strands of endothelial cells at denudation border sloughing from surface. x 1,600

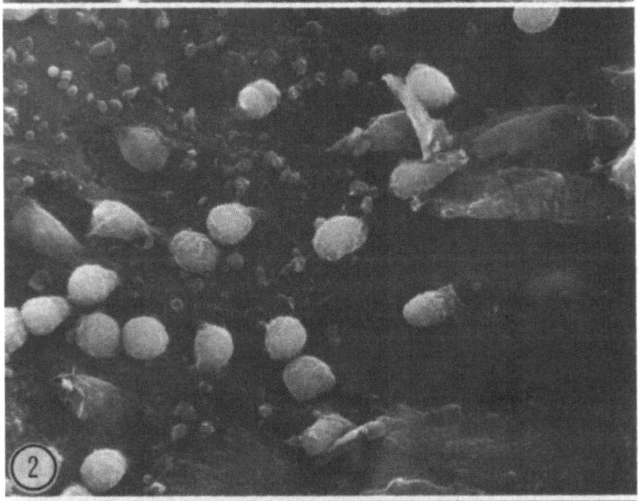

Fig. 2. Regenerating endothelial plate at margin of denudation. Platelets, granulocytes, and lymphocytes adhere to subendothelial space. x 3,100

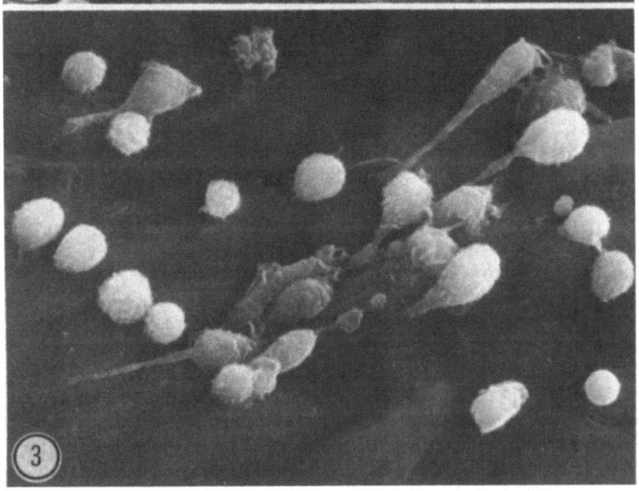

Fig. 3. Cells adherent to endothelium. Heterogeneity of cell size and shape. Lymphocytes are small, globular, and spiculated. Other cells are spindle-shaped with drawn-out pseudopods. x 3,100

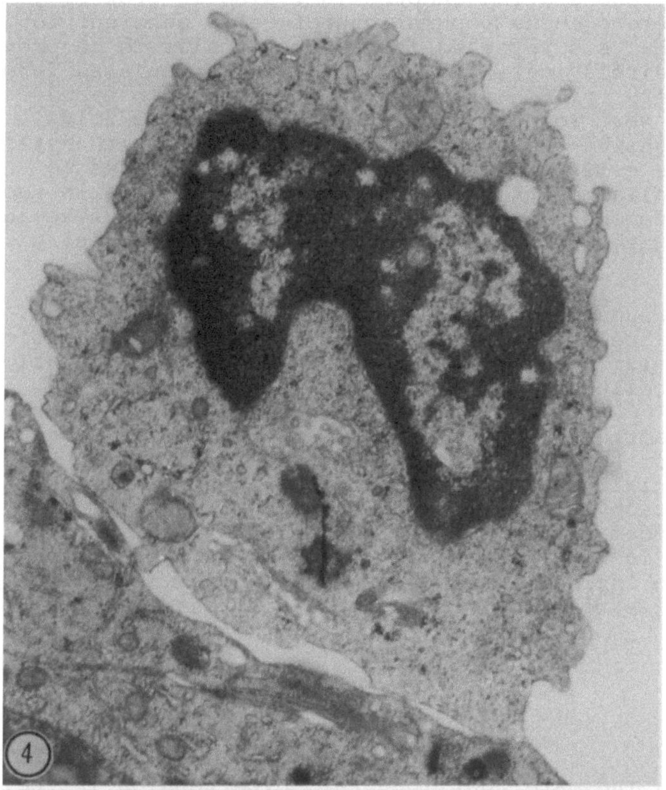

Fig. 4. TEM of adhering small lymphocytic cell with micropinocytic vesicles and fine granular cytoplasm. Centrioles. x 27,200

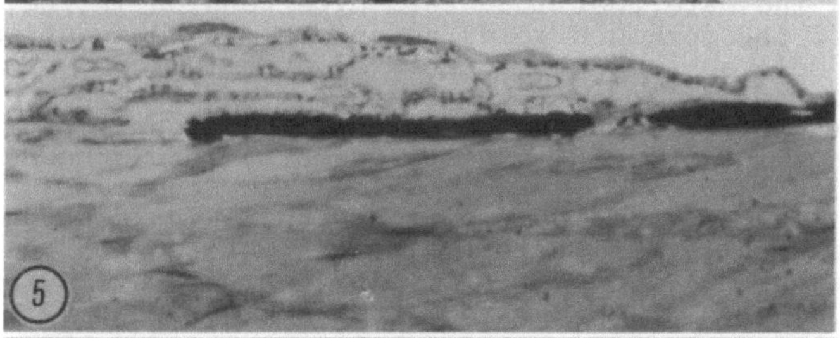

Fig. 5. LM of intimal proliferation, deposition of pericellular matrix material, and breaks in the internal elastic lamina. x 1,250

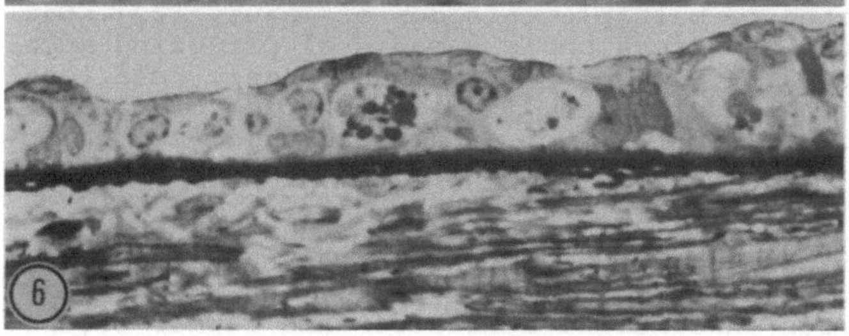

Fig. 6. Inflammatory intimal reaction with plasma cells, large lipid-filled macrophages, and plasma insudation. Internal elastic lamina is intact. x 1,300

points, either in a continuous or in series fashion. These changes were not found in arteries less than 1 mm in diameter. Occasional transverse lesions were also seen. Platelets, lymphocytes, and monocytes adhered to the subendothelial surface (Fig. 2). Regenerative and degenerative endothelial phenomena were noted. Lymphocytes and monocytes were seen beneath regenerating endothelium. Also in proximity to branch points were heterogenous populations of cells adherent to endothelium (Fig. 3). Analysis revealed the adherent cells to be predominantly lymphocytes displaying a small, round, and spiculated form (Fig. 4). Other cells were spindle-shaped with long and thin pseudopods. Resting monocytes were round and exhibited deep coarse folds, while moving monocytes showed the typical triangular shape with hyaloplasmic veils [4]. These cells appear to extend their processes to create breaks in endothelial continuity. The endothelial cells in the area of cell adherence demonstrate heterogeneity of shape and size and do not display the elongated form characteristic of normal coronary artery endothelium.

TEM and LM verify the denudation process, as well as cell adherence, and provide evidence for cell characterization. Breaks and duplication of the internal elastic lamina and intimal smooth muscle proliferation with deposition of pericellular matrix material were seen (Fig. 5).

To study the intimal reaction in areas of cell adherence SEM preparations were reembedded for LM and TEM analysis. Results showed the intimal reaction to be strikingly different from that found in intimal smooth muscle proliferation. Plasma insudation, plasma cells, macrophages, and occasional foam cells dominate the picture (Fig. 6). The internal elastic lamina is retained and the reaction is typically confined to the intima. The intimal reaction is occasionally superimposed upon a basal proliferative reaction with pericellular matrix deposition.

Discussion

The results of our study indicate that long term hypertension in the coronary artery system of the foxhound leads to endothelial denudation, inflammatory reactions, cell adhesion to luminal endothelium, and proliferation in the subendothelial space. The finding of endothelial denudations lends support to the response to injury hypothesis first proposed by Rudolf Virchow in 1856 [29] and subsequently strengthened by numerous investigations [28,25,14]. Recent evidence demonstrates that hypercholesterolemia alone or in combination with a single denudation leading to intimal proliferation is not sufficient to produce the proliferative fibrous plaque considered by many to be the precursor of atherosclerosis [8]. However, atherosclerosis does develop in hyperlipidemic rats subjected to multiple episodes of injury [22]. Our study shows that sustained hypertension is capable of producing such episodes of injury and repair, and suggests a pathogenetic relationship of hypertension to atherosclerosis. The localization of these denudation lesions coincides with that of atherosclerotic lesions in human coronary arteries, namely in the vicinity of major branch points. This gives support to theories implicating a hemodynamic basis for atherosclerosis [17,19]. SEM findings of bright, sloughing, spindle-shaped cells proximal to coronary artery branching points after minimal hypertension are similar to those described by Reidy and Bowyer in their report of spontaneous endothelial denudation and repair occurring in branch points associated with high stress factors [27]. These cells could indeed represent the first stages of focal injury in areas of hemodynamic stress.

The observation of platelets adhering to the subendothelial space is significant. It is known that smooth muscle proliferation is prevented when rabbits are made thrombocytopenic with antiplatelet serum [24].

Spontaneous atherosclerosis in pigs with von Willebrand's disease is significantly reduced [5]. Postoperative vein-graft patency is enhanced by using platelet-inhibitor drugs in coronary bypass operations [9]. There is evidence, then, that platelets may play a role in stimulation of smooth muscle proliferation. Moreover, platelets have been shown to enhance human and rabbit monocyte adherence release of mediators or direct platelet-monocyte binding [26]. This may in part explain the adherence of monocytes in our experimental model of hypertension.

Our results demonstrate that lymphocytes, monocytes, and granulocytes adhere to the endothelial surface of coronary arteries after prolonged hypertension. There is some evidence suggesting that these cells may play a role in the denudation process itself. Endothelial denudation can be mediated by neutrophil-derived proteases [20]. The ability of polymorphonuclear granulocytes to aggregate and damage cultured endothelial cells in vitro when exposed to activated complement (C_{5a}) has been demonstrated [10]. Atheroma cholesterol activates complement and aggregates granulocytes [18]. Immunologic injury to the arterial wall has been implicated in the pathogenesis of atherosclerosis in humans [16,7,21].Indeed, immunologic mechanisms may be important in the role of tobacco as a risk factor for atherosclerosis [2]. Immune complex disease and graft rejection has induced atherosclerosis closely resembling that in humans [1].

Our investigations attempt to isolate the effect of hypertension on coronary artery morphology in dogs. The results show that hypertension leads to subendothelial proliferation, but not to full-blown atherosclerosis. Lesions are localized in areas where hemodynamic factors are thought to play a critical role. The findings support experimental models designed to elucidate the importance of endothelial denudation in atherogenesis. The finding of lymphocytes, monocytes, and granulocytes adhering to endothelium, as well as their incorporation in the intima provides evidence for an inflammatory and/or immunologic component in atherogenesis.

Acknowledgements. This work was supported by the German Research Foundation SFB 90 B/5. We are indepted to the excellent technical assistance of Renate Stahl and Barbara Brühl. Special thanks are offered to Prof. Dr. H.D. Fahimi and Prof. Dr. H. Jellinek for their kind and helpful suggestions.

References

1. Alonso DR, Starek P, Minick CR (1977) Studies on the pathogenesis of atheroarteriosclerosis induced in rabbit cardiac allografts by the synergy of graft rejection and hypercholesterolemia. Am J Pathol 87: 415
2. Becker CG, Dubin T, Weideman HP (1976) Hypersensitivity to tobacco antigen. Proc Natl Acad Sci USA 73: 1712
3. Bell ET, Clawson BJ (1928) Primary (essential) hypertension: A study of 420 cases. Arch Pathol 5: 939
4. Bessis M (1973) Living blood cells and their ultrastructure. Springer Berlin Heidelberg New York, p 305
5. Bowie EJW, Fuster V, Owen CA Jr, Brown AL (1975) Resistance to the development of spontaneous atherosclerosis in pigs with von Willebrand's disease. Fifth Congress of the International Society on Thrombosis and Haemostasis, Paris p 132
6. Breckenridge A, Dollery CT, Parry EH (1970) Prognosis of treated hypertension: Changes in life expectancy and causes of death between 1952 and 1967. Q J Med 39: 411

7. Burch JE (1974) Viruses and arteriosclerosis. Am Heart J 87: 407
8. Clowes AW, Collazzo RE, Karnovsky MJ (1978) A morphologic and permeability study of luminal smooth muscle cells after arterial injury in the rat. Lab Invest 39: 141
9. Chesebro JH, Clements IP, Fuster LR, Elveback LR, Smith HC, Bardsley WT, Frye RL, Holmes DR Jr, Vlietstra RE, Pluth JR, Wallace RB, Puga FJ, Orszulak TA, Piehler JM, Schaff HV, Danielson GK (1982) A platelet-inhibitor-drug trial in coronary artery bypass operations: Benefit of perioperative dipyridamole and aspirin therapy on early postoperative vein-graft patency. N Engl J Med 307: 73
10. Craddock PR, Hammerschmidt DE, Dalmasso AP, White JG, Jacob HS (1977) Complement (C5a) induced granulocyte aggregation in vitro: a possible mechanism of complement-mediated leukostasis and leukopenia. J Clin Invest 60: 260
11. Dawber TR, Kannel WB (1958) An epidemiologic study of heart disease: The Framingham study. Nutr Rev 16: 1
12. Hüttner I, Bandonnel MC, Elemer G, Gabbiani G (1979) Aortic intima of the rat in various phases of hypertension. Ex Mol Pathol 31: 191
13. Hüttner I, Costabella PM, De Chastonay C, Gabbiani G (1982) Volume, surface, and junctions of rat aortic endothelium during experimental hypertension. Lab Invest 46: 489
14. Fishman JA, Ryan GB, Karnovsky MJ (1975) Endothelial regeneration in the rat carotid artery and the significance of endothelial denudation in the pathogenesis of myointimal thickening. Lab Invest 32: 339
15. Forssmann WG, Ito S, Weihe E, Aoki A, Dym M, Fawcett D (1977) An improved perfusion fixation method for the testis. Anat Rec 188: 307
16. Friedman RJ, Moore S, Signal DP (1975) Repeated endothelial injury and induction at atherosclerosis in normolipemic rabbits by human serum. Lab Invest 30: 404
17. Fry DL (1969) Certain chemorheologic considerations regarding the blood vascular interface with particular reference to coronary artery disease. Circulation Suppl 4: 38
18. Greenberg CS, Hammerschmidt DE, Craddock PR, Jacob HS (1979) Atheroma cholesterol activates complement and aggregates granulocytes: Possible role in ischemic manifestations of atherosclerosis. Trans AAP 92: 130
19. Gutstein WH, Farrel GA, Armellini C (1973) Blood flow disturbance and endothelial cell injury in preatherosclerotic swine. Lab Invest 29: 134
20. Harlan JM, Killen PD, Harker LA, Striker GE (1980) Endothelial denudation mediated by neutrophil-derived neutral proteases. Clin Res 28: 547A
21. Howard AN, Patelski J, Bowyer DE, Gresham GA (1971) Atherosclerosis induced in hypercholesterolemic baboon by immunologic injury, and the effects of intravenous polyunsaturated phosphatidyl choline. Atherosclerosis 14: 17
22. Koletsky S, Snaydar RM (1981) Atherosclerosis following balloon catheter injury to the carotid artery and the aorta of hypertensive rats with normolipidemia or hyperlipidemia. Am J Pathol 103: 105
23. Mittmann U, Brückner UB, Keller HE, Kohler U, Vetter H, Waag KL (1980) Myocardial flow reserve in experimental cardiac hypertrophy. Basic Res Cardiol 75: 199
24. Moore S, Friedman RJ, Singal DP, Gauldie J, Blajchman MA, Roberts RS (1976) Inhibition of injury induced thromboatherosclerotic lesions by antiplatelet serum in rabbits. Thromb Haemostas 35: 70
25. Moore S (1979) Endothelial injury and atherosclerosis. Exp Mol Pathol 31: 182
26. Musson RA, Henson PM (1979) Humoral and formed elements of blood modulate the response of peripheral blood monocytes. I. Plasma and serum inhibit and platelets enhance monocyte adherence. J Immunol 122: 2026
27. Reidy MA, Bowyer DE (1979) Scanning electron microscopy of arteries: the morphology of aortic endothelium in haemodynamically stressed areas associated with branches. Atherosclerosis 26: 181
28. Ross R, Glomset J, Harker L (1977) Response to injury and atherogenesis. Am J Pathol 86: 675
29. Virchow R (1856) Gesammelte Abhandlungen zur Wissenschaftlichen Medicin. Meidinger, Frankfurt, p 499

Discussion

CARO: What was the blood pressure?

MEAIRS: Prestenotic pressures and pressure gradients were determined after three months and after one year of supravalvular stenosis. After 3 months the mean systolic pressure gradient was 31 mmHg with a prestenotic systolic pressure of 134 mmHg and a diastolic pressure of 83 mmHg. After 1 year the mean systolic pressure was about 180 mmHg with a diastolic pressure of 100 mmHg. Foxhounds were perfusion fixed at 100 mmHg which consistently yielded a pressure of 60 mmHg in the ascending aorta proximal to the coronary ostia. Flow was regulated according to heart size in hypertensive dogs.

FRIEDMAN: I was particularly interested in those longitudinal denudations between the ostia of successive branches. You remarked that they appeared to be lined up with the blood flow, and I would just add a caution there that, although they certainly are more lined up than the transverse denudations were, as Dr. Seed and Dr. Nerem have shown, the flow directions at the wall in the coronary arteries may be quite different from the longitudinal direction. The question is, for that particular one you showed which went from branch to branch, from ostium to ostium, did you see any that might have been part way along so that you could tell whether they progressed distally or proximally from an ostium?

Meairs: I have frequently observed a series of denudations extending as a channel between branch points. Unfortunately, I have been unable to ascertain in which direction the denudations progress.

Response of Vascular Endothelium to Unsteady Fluid Shear Stress in Vitro[1]

C. F. Dewey, Jr., M. A. Gimbrone, Jr., S. R. Bussolari, G. E. White, and P. F. Davies

Introduction

We have subjected vascular endothelium to unsteady fluid shear stress. In these experiments, the stress state has been changed discontinuously from one level to another. This has altered the cell morphology and increased the rate of fluid endocytosis. These changes occur even though there is no change in the mitotic rate or the number of cells per unit area. In this paper, we report the details of these experiments and compare the response of the cells to changes in shear with their response to other types of alterations. A concluding section summarizes our understanding of the several time scales governing the morphological and functional changes of endothelium subjected to unsteady shear stress.

Apparatus and Experimental Procedure

Confluent monolayers were prepared from bovine aortic endothelial cells according to methods described elsewhere (Gimbrone 1976). After the cultures had grown to confluence on 12 mm diameter glass coverslips under static conditions, specimens to be subjected to fluid shear stress were transferred to a cone-plate test apparatus (Dewey et al. 1981; Bussolari et al. 1982) that is capable of producing controlled fluid shear stress on the endothelial layer. Although the apparatus is capable of producing a wide range of laminar and turbulent shear stresses, only laminar shear stresses between 1 and 15 dynes/cm^2 were used in the current experiments.

We have used the uptake by the cells of the soluble enzyme tracer molecule horseradish peroxidase (HRP) to measure fluid endocytosis. The procedure, described by Davies et al. (1981), involves incubation of the cells with HRP-containing medium followed by assay to determine the amount of HRP internalized by the cells during the incubation period.

Time-lapse cinematography was used to study the dynamics of the changes in cell shape following a change in shear stress. The method (White et al. 1982) captures a sequence of still photographs of the cells, as seen through a microscope, on 16 mm movie film. The frames are taken at 24 second intervals, so that the resulting movie displays cell motion at approximately 400 times normal speed. The motion of individual cells may be clearly seen. Because of technical difficulties, this procedure has only been used to examine the relaxation of cells under static conditions.

---

1 This research was supported in part by the National Heart, Lung, and Blood Institute, Grants HL 25536 and HL 22602; and by the Whitaker Health Sciences Fund.

Fluid Dynamics as a Localizing Factor for Atherosclerosis
Editors: G. Schettler et al.

In order to quantitate the morphology of cells subjected to shear, a computer program was developed to analyze a video image of cells (Milich 1982). Either a black-and-white photograph or a manual tracing from a photograph could be used as raw data. The program used an edge-detection algorithm to determine the cell boundary; an ellipse having the same total area was then fit to measured boundary using an eccentricity and axis orientation that minimized the deviation between the ellipse and the cell outline. The two primary parameters obtained from this procedure for each cell were the eccentricity and the angle of the cell relative to the direction of flow.

Results and Discussion

When a confluent endothelial monolayer is subjected to a shear stress of 8 dynes/cm², a value comparable to that experienced in vivo, the cells align with the flow direction in something less than 48 hours (Dewey et al. 1981). The degree of elongation in the flow direction is quite striking. The majority of cells have their longest axis aligned within 15° of the flow direction; and the mean degree of eccentricity (measured as the ratio of major axis of the best-fit ellipse representing the cell shape) is 3.2. These shape changes are not accompanied by a change of cell density or a change in mitotic rate; the new morphology occurs by rearrangement and reorientation of existing cells.

The design of our shear stress apparatus does not allow continuous observation of the cell layer during orientation. However, we obtained continuous records of the reverse process: namely, relaxation of a shear-oriented monolayer under static (no-shear) conditions. Fig. 1 presents three selected frames from a time-lapse movie film of a layer, incubated at 8 dynes/cm², relaxing to a randomly-oriented morphology typical of cultured cells in static medium. The mean eccentricity has decreased to a value intermediate to the sheared and unstressed conditions in approximately 8 hours[2].

The individual cells were seen to migrate several times their own diameter in a relatively short time (tens of minutes). Then cells or groups of cells would be quiescent for extended periods (many tens of minutes). Occasionally, a cell would undergo surface speculation and oscillations of the cytoplasm with a period of about one minute while remaining attached to a single site. The frequency of cell motion decreased dramatically as rearrangement became complete, approaching the quiesence of the original static cultures after 24 hours.

Our studies of the time course of cell morphology following a change in stress state have been paralleled with experiments on fluid endocytosis. Fig. 2 presents one interesting set of results. In the two experiments (no stress and 8 dynes/cm²) where the endothelium had been accomodated to its stress condition for at least 48 hours, the endocytosis rates were equal - independent of the absolute level of stress. A change in the stress level, either from zero to 8 dynes/cm² or from 8 dynes/cm² to zero, produced an endocytosis rate that was double that of an accomodated layer.

---

2 A more quantitative description of the relaxation awaits further video analysis of large samples of cells.

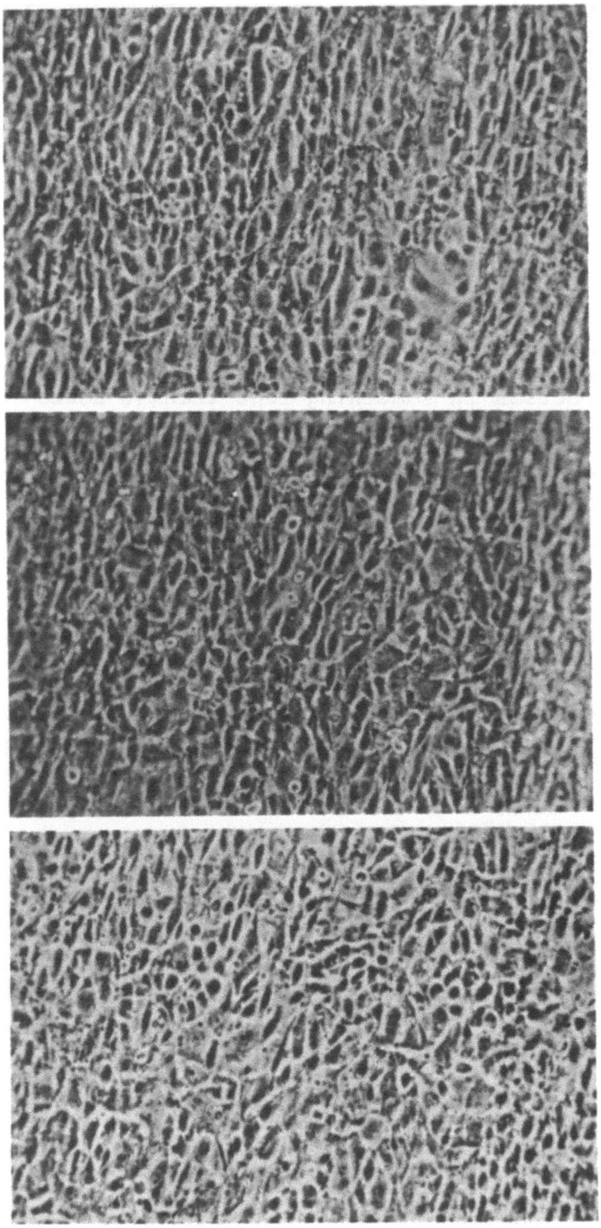

Fig. 1

In a companion experiment, we exposed a confluent specimen of endothelium incubated under static conditions to an oscillating shear; the shear level was alternated between 1 and 8 dynes/cm² every 15 minutes during a two-hour period. Again, the endocytosis rate was twice that of static controls. It should be noted that 1 dyne/cm² is not sufficient to produce endothelial reorientation in our system (Dewey et al. 1981).

SHEAR STRESS : TRANSIENT EFFECT

Fig. 2

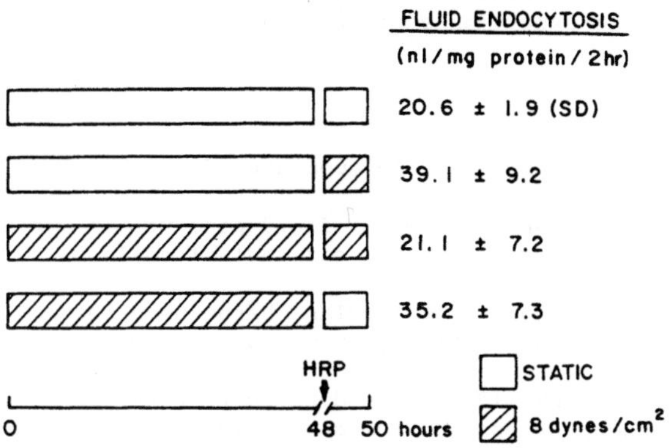

In the endocytosis experiments, shear-induced entry of endothelial cells into the cell cycle was monitored by ^{3}H-thymidine autoradiography for DNA synthesis. In no instance did subsequent exposure to shear (up to 15 dynes/cm^2) result in DNA synthesis during a subsequent 24 hour period. Therefore, changes in endocytosis rate under shear were not attributable to fundamental cell cycle changes which have been demonstrated under conditions of defined growth (Davies et al. 1980).

Concluding Remarks

There now exists evidence from a variety of sources that fluid shear stress plays an integral role in endothelial morphology and function. An important conclusion from the experiments reported here is that, for certain functional processes such as endocytosis , a *change* of stress may be as important as absolute stress *level* in determining cell response.

Clearly our present experiments have only begun to delineate the dynamics of endothelial response to unsteady forces. The cells appear to reach a quiescent accomodation with a steady state of stress in times of the order of 24 hours. This time may depend upon the sub-cellular matrix as has been observed during regeneration (Jackman 1982; Madri and Stenn 1982). We have noted that elevations of endocytosis occurred when the shear stress was elevated for only 50% of the time. Additional experiments are necessary to determine if cells can accomodate to a continuously oscillating stress in the same manner that they accomodate to a new steady level of stress. And, if an accomodation is reached, what is the range of oscillation magnitudes and oscillation frequencies over which this accomodation occurs?

Finally, we note that endothelium in vivo is subjected to oscillating shear stress as a natural condition. The dominant frequency of oscillation is, of course, the heart frequency, although the waveform of wall shear stress contains significant harmonic components of 4 times the heart frequency. The magnitude of the shear can vary by a factor of 4 times during a single heartbeat. On a longer time scale (matters of minutes to hours) exercise can effect an increase of a factor of 4 in cardiac output and a similar increase in wall shear stress. And in certain cases such as constricted arteries, turbulence can be produced. Here, there are additional high-frequency (50-500 Hz) fluctuations in shear of smaller magnitude superimposed upon the existing shear patterns.

Acknowledgements. We are grateful to Dr. K. Fujiwara who was instrumental in establishing the time-lapse cinematography procedures; and to Ms. E. Shefton for her invaluable assistance in all phases of the experiments reported here.

References

Bussolari SR, Gimbrone MA, Dewey CF (to be published) An apparatus for subjecting living cells to fluid shear stress.

Davies PF, Selden SC, Schwartz SM (1980) Enhanced rates of fluid pinocytosis during exponential growth and regeneration by cultured arterial endothelial cells. J Cell Physiol 102: 119-127

Davies PF, Rennke HG, Cotran RS (1981) Influence of molecular charge upon the endocytosis and intracellular fate of peroxidase activity in cultured arterial endothelium. J Cell Science 49: 69-86

Dewey CF, Bussolari SR, Gimbrone MA, Davies PF (1981) The dynamic response of vascular endothelial cells to fluid shear stress. J Biomech Eng 103: 177-185

Fishman JA, Ryan GB, Karnovsky MJ (1975) Endothelial regeneration in the rat carotid artery and the significance of endothelial denudation in the pathogenesis of myointimal thickening. Lab Invest 32: 339-351

Gimbrone MA (1976) Progress in hemostasis and thrombosis, vol 3. Grune and Stratton New York, pp 1-28

Jackman RW (1982) Persistence of axial orientation cues in regenerating intima of cultured aortic explants. Nature 296: 80-83

Madri JA, Stenn KS (1982) Aortic endothelial cell migration. I. Matrix requirements and composition. Am J Path 106: 180-186

Milich D (1982) Quantitative analysis of the orientation of endothelial cells subjected to a fluid shear stress. S.B. Thesis, M.I.T., Cambridge, MA, USA

Reidy MA, Sxhwartz SM (1981) Endothelial regeneration. III. Time course of intimal changes after small defined injury to the rat aortic endothelium. Lab Invest 44: 301-308

Schwartz SM, Haudenschild CC, Eddy EM (1978) Endothelial regeneration I. Quantitative analysis of initial stages of endothelial regeneration in rat aortic intima. Lab Invest 38: 568-580

Sholley M, Gimbrone MA, Cotran RS (1977) Cellular migration and replication in endothelial regeneration: A study using irradiated endothelial cultures. Lab Invest 36: 18-25

White GE, Fujiware K, Shefton EJ, Dewey CF, Gimbrone MA (1982) Fluid shear stress influences cell shape and cytoskeletal organization in cultured vascular endothelium.

Discussion

STEHBENS: In the endothelial preparations with the shear stress, was there any evidence that the position of the nucleus in relationship to the length or axis of the cell varies with shear? It appears to in the rabbit aortic endothelium.

DEWEY: We have not quantitated the degree of nuclear orientation as a continuous function of shear stress level. At 8 dynes per square centimeter, which has been one of our typical levels that we run, if you fit an ellipse to the cells on the average, the eccentricity would be about 3.5; that means the cells are 3.5 times as long as they are wide. At 5 dynes per square centimeter, the degree of orientation depends in part upon the degree of confluence of the layer, and in fact, if it is a fully confluent layer, below 5 dynes per square centimeter you do not see any rearrangement at all. So there is a certain shear threshold that undoubtedly depends upon the details of attachment. We intend to go up to higher stresses, but we simply have not done that in a quantitative manner. I think the degree of alignment is a highly non-linear function of shear stress. A cell is either aligned or it is not aligned; and when it is aligned the eccentricity of the apparent ellipse does not change that much.

SCHMID-SCHÖNBEIN: I was interested in your cells which started "dancing". Did you notice that they reached more into the fluid layer? Did you actually watch them? Could you see that, because if they did, this would offer a good explanation of how these cells can be oriented in shear?

DEWEY: It is very hard to tell because this was a sort of an en face view with the light microscope. It is possible that they came up above the ambient surface levels somewhat, but without stereoscopic facilities I could not tell.

C. SCHWARTZ: You may not have done the experiment, but I would like to know if you have or contemplate it. If you have reorientation under shear and then remove the cells, do they produce an extracellular matrix which would be recognized by another group of cells which would provide orientation along the same lines as the cells after shear?

DEWEY: You have read our grant renewal request.

Injury at the Vascular Surface

S. M. Schwartz, M. A. Reidy, and G. K. Hansson

Introduction

The last ten years have seen an exponential growth in interest in the role of the endothelium in atherosclerosis and in the idea that suggests that the cell affected by shear, the endothelial cell, may play a critical role in lesion formation. This interest has come from two quite different sources. First, there is the apparent coincidence of areas of high shear with areas with a propensity to develop atherosclerosis. Second, there is the biochemical evidence that the endothelial cell is intensively active in a number of metabolic functions capable of modulating lesion formation. All of this has tended to blur our lack of a definition for endothelial injury. For example, Ross' and Harker's (1976) discovery of the platelet-derived growth factor (PDGF) led to a major emphasis on denuding injuries as a site where platelets could interact with the vessel wall and stimulate smooth muscle proliferation. Actual studies of denuding injuries, however, have been very limited except over advanced lesions or in response to the balloon catheter. Studies of hypertensive or hyperlipemic animals have not shown denudations unless overt atherosclerotic lesions were already present (Schwartz et al. 1980; Hansson and Bondjers 1980). Absence of evidence for denudation could be due to technical difficulties - a small area of denudation is unlikely to be detected in microscopic examination of cross sections of the vessel wall. Scanning electron microscopy (SEM), however, provides rapid and thorough examinations of large areas of vascular surfaces in great detail. Many square meters of the endothelial surfaces of rabbits, rats, monkeys, and other species have been examined in the SEM. These studies, in both normal and hyperlipemic animals, have shown a continuous layer of endothelial cells covering the surface, with no evidence of denudation until after lesion formation (Davies and Bowyer 1975; Clark and Glagov 1976; Bylock et al. 1977; Reidy and Schwartz 1980). These data indicate that denudation either does not occur in the unmanipulated animal, or involves areas that are too small to be detected by SEM.

In this brief review we would like to reconsider current concepts of denudation in terms of the time course of atherosclerosis, with particular emphasis on the sort of initial changes in the endothelium that might reflect endothelial injury due to flow.

1. Natural History of Atherosclerosis

The issue of time course is an important one to consider. If we are to regard flow as an independent risk factor, we need to consider its effects on an uninjured vessel wall. Figure 1 represents a reasonable sequence of events based on the concepts reviewed here. This scheme does not assume that endothelial injury is the initial event in atherogenesis, but it does include endothelial denudation as a causal event, occurring at some time during lesion formation. There is evidence that

Fluid Dynamics as a Localizing Factor for Atherosclerosis
Editors: G. Schettler et al.

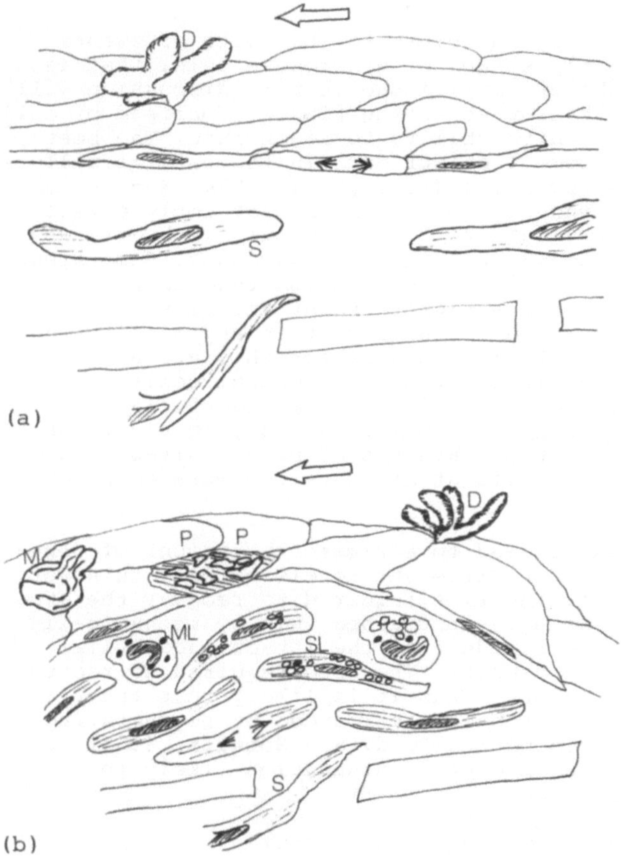

Fig. 1. (a) *Normal maintenance of integrity.* Under normal conditions the endothelium maintains itself as a continuous cell layer despite spontaneous cell death and cell turnover. The earliest recognizable atherosclerotic lesions can be found in the presence of the intact endothelium. (b) *Loss of integrity.* As lesions progress, the monolayer loses its ability to maintain continuity. At this point it is likely that thrombosis results in rapid progression of the lesion. The possibility of intervening in this step is currently of considerable clinical interest, although we do not know why lesion progression leads to loss of continuity. D = dying endothelial cell; S = smooth muscle cell; M = monocyte/macrophage; P = platelet; ML = macrophage filled with lipid; SL = smooth muscle cell filled with lipid

a sequence similar to this does occur. Both the Velicans' and Thomas' group have described intimal accumulations of smooth muscle cells preceding any more obvious lesions (Velican and Velican 1979; Thomas et al. 1981). These sites appear to be the ground in which lesions develop. Their origin is unclear, but the association of "normal" intimal thickening with "pathologic" lesion formation has been recognized for over a decade (French 1967). Similarly, there is no doubt that thrombosis occurs as lesions develop (Chandler and Pope 1975) and, given Ross' observations of PDGF, it seems highly likely that this event will accelerate lesion progression, perhaps playing a critical role in the eventual formation of occlusive lesions (Ross and Harker 1976).

2. Nondenuding Desquamation

Against this background we can again return to the sorts of events likely to occur during the early stages of atherosclerosis, that is, in the normal vessel wall. The critical issue in this discussion will be the treatment of the endothelium as living tissue comprised of cells with finite life spans and a finite ability to regenerate. The best examples of these properties come from our data in the rat, as well as studies by others in rabbits and swine (Bell et al. 1974; Kunz et al. 1978; Thomas et al. 1981). Turnover studies in the rat imply a rate of cell loss of approximately $1 \cdot 10^{-2}$ to $2 \cdot 10^{-3}$ per day (Schwartz and Benditt 1973, 1976). This turnover, however, is nonrandom. Some areas of the vessel show much higher rates of replication than others. For example, although one might expect a high turnover due to shear, turnover at the aortic trifurcation is the lowest at any site in the aorta (Schwartz and Benditt 1976). The reasons for this are not obvious, but it is reasonable to imagine that endothelial cells adapt to shear. It may also be that the effects of shear on endothelial turnover are highly focal. Kunz and his co-workers (1978), in a very detailed study on distribution of replication near branch sites, found an increase in replication when cells at branch sites were compared with the adjacent wall.

These values for turnover can be used to estimate the amount of denudation we might expect to find for different levels of cell turnover. One additional value is needed: the time required to recover the area exposed when a cell is lost. We have approached this by mechanically abrading strips of cells one cell wide from the intact endothelial lining of the aorta of rats and rabbits. The time required to recover a strip two cells wide is 6-8 hours. Thus it is likely that the time needed to replace a single cell, totally surrounded by regenerating neighbors, is no more than three hours (Reidy and Schwartz 1981). On this basis, we can calculate the amount of denudation likely to be present at sites of turnover:

(1) $E = A \times R \times T$

E = denuded area = (μm^2/cell)
A = area of each cell ~ 10^3 μm^2
R = rate of turnover
T = time required to replace each cell ~ 3 hours

As you can see, even for relatively high rates of turnover, the average exposed area is quite small. Furthermore, the probability of our seeing a separation between two cells depends not only on the average area of exposed surface, but on the size of the cells. Since endothelial cells have an average circumference (C) of 200 microns, we can divide E by the value 1/2 C to estimate the likely distance between cells at sites of turnover:

(2) $D = (ART)/C/2$

D = average distance separating regenerating cells
C = circumference of an endothelial cell

This means that the average distance between cells turning over at 10^{-3} cells per day would be no more than one nanometer, obviously well below our current abilities to detect denudation. Even in an area of high turnover, $T = 10^{-1}$, the average separation should be no more than 0.1 microns.

The point of this exercise is that substantial but transient amounts of denudation could exist without being detectable by SEM. Such areas, if occurring repeatedly, might result in significant local alterations in permeability or even in interactions with platelets. The fact that the thrombogenecity of freshly exposed, denuded surfaces lasts only a few hours (Mustard et al. 1981; Goldberg et al. 1980) suggests that such areas of denudation, if they existed, might play a major role in the interactions of the vessel wall and platelet. This, however, is all speculation. The estimates are for the *maximal* extent of endothelial denudation likely to be present in vivo. As yet, there is no documentation of the existence of injuries of this sort, prior to the appearance of overt lesions.

One approach to this question is to look at the events associated with increased desquamation. There is substantial evidence for increases in the turnover of endothelial cells in response to risk factors. Hypertension causes an increased uptake of thymidine in the rat aortic endothelium (Schwartz and Benditt 1977). In experimental hypercholesterolemia there is an increased thymidine labeling of the endothelium on atherosclerotic lesions (Hansson and Bondjers 1980). This increase in DNA synthesis may precede the development of the lesions (Florentin et al. 1969). The localization of normal cell turnover is also consistent with a possible etiologic role in atherosclerosis. ^{3}HTdR-labeled cells predominantly appear in clusters at branching points, both in normal and hypertensive animals (Wright 1972; Schwartz and Benditt 1977; Kunz et al. 1978). This clustering may be related to increased cell turnover at sites of hemodynamic strain in these regions (Fry 1968). Under experimental conditions endothelial cells can be detached from perfused walls by shearing forces in excess of 200 dynes/cm^2 (Fry 1968). Such forces can occur in vivo and if single cells or small patches of cells are removed, the endothelium has the capacity to re-establish continuity rapidly (Reidy and Schwartz 1981). In the normal rat such small areas of desquamation are covered by the migration of surrounding cells, without proliferation, in a few hours. We simply do not know whether prolonged normal turnover at single sites or the changes in turnover seen in hypertension and hyperlipemia may not ultimately lead to a loss of endothelial continuity.

Endotoxemia provides the one well-documented form of desquamating endothelial injury (Gaynor 1971; Reidy and Bowyer 1977). Our studies, however, show that turnover increases without formation of denuded areas. The presence of cells loosely attached to the monolayer implies that this lack of denudation is a result of the coordinated undermining of detaching cells by adjacent viable endothelial cells. At least in this case, we appear to have a form of "nondenuding desquamation" (Reidy and Schwartz 1982).

As discussed above, areas of nondenuding desquamation might show transient exposure of the subendothelium. Moreover, even when the endothelium is otherwise continuous, it is very likely that areas of nondenuding desquamation would show altered permeability to plasma proteins, since junctions between regenerating endothelial cells are poorly formed (Spangoli et al. 1981; Schwartz et al. 1975). Finally, the endothelium at sites of repeated turnover might eventually lose the ability to replicate, much as happens in cell cultures when endothelial cells undergo multiple passages (Schwartz 1978).

3. Nondenuding Injury

So far we have only considered the effect of denuding injuries. Considering all the new information on the metabolic functions of endothelial cells, it seems certain that endothelial cell dysfunction and death have many important metabolic effects (Ross and Schwartz 1982). These metabolic changes, however, suffer from a problem quite different from the subjective problem involved in evaluating ultrastructural evidence for cell injury. While these are objective changes, they have largely been described in vitro. It is difficult to generalize these phenomena to in vivo, particularly when the focal nature of atherosclerotic lesions suggest that any relevant injury in the endothelium is also likely to be focal.

In contrast, morphologic data can tell us about focal injuries that occur in vivo. Ultrastructural evidence for more subtle forms of endothelial cell injury has been reported in many conditions known to induce atherosclerosis (Asmussen and Kjeldsen 1975; Bylock et al. 1977; Svendsen 1979). Weber et al. (1974) described changes in the thickness of the cells' luminal coat. Despite these morphologic data, our knowledge of endothelial injury remains uncertain for four reasons. First, the functional significance of changes seen in the electron microscope is often unclear. Second, in most cases these are subjective changes, and it is difficult to reproduce the data. Third, as we have discussed, the evidence that endothelial injury initiates atherosclerosis by denudation is not convincing. Finally, we need to remember that ultrastructural changes occur long after irreversible cell injury. For example, cardiac monocytes persist for days after cell injury, even in areas where perfusion by blood is restored (Jennings et al. 1975). As we will see, this may be an important issue in the endothelium as well. The one category of injury that can be studied in vitro and can be unequivocally identified in intact tissue is cell death. We would like to present evidence for the occurrence of dead cells in the continuous endothelium.

4. *Mechanisms of Cell Death*

It is important to begin by considering the current criteria for defining cell death. Of key importance is the identification of "the point of no return"; i.e., the transition from reversible to irreversible cell injury.

One hypothesis has been that the "point of no return" reflects the increased permeability of the lysosomal membrane, with release of lysosomal hydrolases into the cytosol (deDuve 1964). However, later studies have shown that lysosomes do not rupture until very late in the progress of cell injury - far beyond the point of no return (Hawkins et al. 1972; Ericsson and Brunk 1977; Kane et al. 1980). Another candidate is the loss of mitochondrial energy production to a level where the membrane pumps fail to maintain gradients. This generates intracellular Ca^{++} ions that are incompatible with the functions of Ca^{++}-regulated enzymatic processes in the cell, and irreversible cell death (Trump and Arstilla 1975; Laiho and Trump 1977).

This mitochondrion-centered hypothesis has been challenged by John Farber and his co-workers (1981). In studies of ischemia in the rat liver, they have found that ischemic periods that lead to irreversible cell injury also lead to irreversible loss of mitochondrial functions. If Ca^{++} accumulation in the mitochondria is prevented by chlorpromazine, however, the mitochondrial dysfunction becomes reversible (Chien et al. 1977; Mittnacht et al. 1979). This suggests that mitochondrial dysfunction is a response to an earlier, irreversible inflow of Ca^{++} through the plasma membrane (Jennings et al. 1975). In our own studies on cell death in vascular endothelial cells, we have used the calcium ionophore A23187 for inducing diffuse cell death of cultured bovine aortic endothelial cells (Hansson and Schwartz 1982b). If aortic endothelial cells are incubated with A23187 and extracellular calcium at a physiological concentration, the viability of the cells is reduced with kinetics that are both time- and dose-dependent. No such alterations are observed in the absence of extracellular calcium. In summary, the identification of the point of no return in cell death is still under debate. Many data point to a crucial role of Ca^{++} in this process, and there is no doubt that a large accumulation of Ca^{++} represents an event occurring after a cell has become irreversibly injured.

5. Cell Death in Situ

To this point our discussion of cell death suffers from the same defects as the literature on cell injury, as discussed earlier. The ultrastructural evidence is not quantitative, and the in vitro data may not represent an in vivo phenomenon. Even if we had a marker visible by light microscopy, the small number of cells seen in section would result in unsatisfactory, small sample sizes (Hugod et al. 1978). Transmission electron microscopy (TEM) is no better. The small area seen at high magnification gives precise information about the state of the single cell, but the ability of such samples of single cells to be representative can always be questioned. SEM permits evaluation of a large number of cells, and should be ideal for detecting denudation. Studies on cell injury, however, are hampered by the fact that intracellular changes are not reflected in the morphology of the cell surface until late in the injury sequence. Furthermore, the sequence itself is not as clearly defined for SEM as for TEM.

The most promising approach is the use of cytochemical markers evident in en face preparations. This allows one to sample the entire vascular surface. For example, the frequencies of replicating endothelial cells is as low as $1 \cdot 10^{-3}$ (Schwartz et al. 1980). As we have already discussed, this is a measure of cell turnover. Since viable cells may detach, it is not a measure of the frequency of cell death. Björkerud and Bondjers (1972) used Evans blue, trypan blue and nigrosin for studying endothelial cell injury. They showed that there are injured endothelial cells in the normal rabbit aorta, and that such cells have a characteristic distribution similar to the distribution of cell replication, with an increased frequency around orifices. Problems with this technique are related to the diffusion of the dye to the subendothelium, and to the risk of inducing artifacts by mechanical force during the staining.

Dr. Hansson developed an immunocytochemical technique for detection of endothelial cell injury as part of his thesis project at the University of Göteborg (Hansson et al. 1982). When the plasma membrane ceases to function as a barrier, IgG penetrates into the cell and is bound to cytoskeletal structures (Hansson et al. 1979, 1980). The specificity of the technique has been established by comparing it with dye exclusion tests (Hansson et al. 1979) and with ultrastructural criteria of cell injury (Hansson et al. 1980). We have developed a quantitative measurement of this change using en face preparations of aortic endothelium. IgG is demonstrated in injured cells of such preparations by an immunoperoxidase or immunofluorescent technique, and the frequency of IgG-positive cells is an inverted estimate of the viability of the cell population. More than 85% of these IgG-positive cells also contain Ca^{++} deposits, as indicated by the accumulation of the calcium probe, chlorotetracycline (Hansson et al. 1982). This implies that the IgG-positive cells have passed the point of no return and are therefore destined to undergo cell death and necrosis.

The immunocytochemical technique has made it possible to study cell death in the rat aortic endothelium. 2.8% of the cells in the thoracic aorta of three-month-old rats were dead according to the IgG technique (Hansson et al. 1982). Earlier studies of ^3H-thymidine labeling have indicated a replication rate of 1% in such rats (Schwartz and Benditt 1977). Taken together, these data suggest that dead cells may remain in the endothelium for several days.

We have turned to an in vitro system to document the fate of cells with these properties. These studies used confluent cultures of bovine aortic endothelium. After a pulse label with trypan blue, the average residence time of a positive cell was three days. The same cells could be shown to bind IgG present in the medium and to stain with chlortetracycline. Direct time-lapse video microscopic observations showed a prolonged process in intracellular movements, gradual fragmentation and undergrowth by surrounding normal cells (Hansson and Schwartz 1982a). The entire process takes about three days. During that time, much as we proposed from studies in vivo, the surrounding cells undergrow the dying cells and maintain a continuous cell sheet (Reidy and Schwartz 1982). These data indicate that irreversibly injured cells can remain in the otherwise normal endothelium for hours and days. It is possible that they can be of pathophysiologic significance during this time.

6. Significance of Cell Death in Situ

At this point the significance of endothelial cell death in situ is open to speculation. There are some changes, particularly in permeability, that seem certain to be associated with focal dead cells. It is interesting to speculate whether this accounts for the focal changes in permeability seen in hyperlipemic rabbits by Stemerman (1981). These changes occurred in the absence of any evidence for focal denudation by SEM. This leads us to the distinction between cell death and loss of cells from the endothelium. Necrosis, defined as the breakdown of cell structures, does not occur until many hours after cells have become irreversibly injured, i.e. dead. There is no comparable knowledge about the line between "cell death" and cell detachment, so it is reasonable to believe that "dead" cells might remain as part of the continuous endothelium for some time. We have begun a collaborative work with Dr. Hansson to estimate the frequency of cell death by applying his technique to Häutchen preparations. We can use this to calculate an approximate residence time, t_r, for dead cells:

f = frequency of dead cells = 1%

$$t_r = \frac{f}{r}$$

r = cell turnover rate = 10^{-3}/day

If we assume that *all* the turnover begins with death, then:

t_r = approximately 10 days

This seems too long. Turnover of injured or normal cells may never be detectable as dead cells. This would mean that the real value of t_r is even longer. On the other hand, Dr. Hansson's technique may overestimate "f" due to some as yet unsuspected artifact. In any case, these data point out the possibility that substantial numbers of dead cells may have significant alterations in the properties likely to influence development of atherosclerotic lesions. It is worth considering the possible metabolic interactions between the process of cell death and the processes associated with atherosclerosis. For example, Linder (1981) suggested that cytoskeletal proteins of dead cells may bind and activate complement. Monocytosis is a prominent feature of the formation of atherosclerotic lesions in several species (Fowler et al. 1979; Gerrity et al. 1979; Hansson and Bondjers 1980). Recent studies by one of us (Göran Hansson), have indicated that monocytes preferentially bind to areas with endothelial injury in the isolated and perfused rabbit aorta. This binding is substantially reduced by preincubating the aorta with Protein A, suggesting that an interaction between IgG in these areas and the Fc receptors of the monocytes promote this binding (Hansson et al. 1981).

Further exploration of such phenomena and studies of endothelial cell death in relation to established risk factors for atherogenesis are presently underway in our laboratories.

7. Biology of Shear

Perhaps the most important issue raised by this article, in the context of a textbook devoted to flow, is our need to better define the interface between the physical phenomenon of blood/wall interactions and the biology of vessel wall cells. From the standpoint of the rheologist, we badly need more information from the rheologist about the sorts and magnitudes of forces impinging on the endothelial cell. In turn, the rheologist needs to have a better idea of the range of changes an endothelial cell may manifest at sites of flow, particularly in vivo. In that range, cell death is clearly the most extreme form of injury. More subtle injuries can be, and have been, detected in vitro, but these phenomena are difficult to document in the intact animal. Nonetheless, the issue and the challenge it presents are clear.

This work was supported by NIH grants HL26405 and HL03174, American Heart Association grant 81-837 and a grant from R.J. Reynolds Industries, Inc. G.K.H. is supported on an NIH Fogarty International Fellowship. S.M.S. is an Established Investigator for the American Heart Association.

References

Assmussen I, Kjeldsen K (1975) Intimal ultrastructure of umbilical arteries. Observations on arteries from newborn children of smoking and nonsmoking mothers. Circ Res 36: 579-589

Bell FP, Gallus AS, Schwartz CJ (1974) Focal and regional patterns of uptake and the transmural distribution of ^{131}I-fibrinogen in the pig aorta in vivo. Exp Mol Pathol 20: 281-292

Björkerud S, Bondjers G (1972) Endothelial integrity and viability in the aorta of the normal rabbit and rat as evaluated with dye exclusion tests and interference contrast microscopy. Atherosclerosis 15: 285-300

Bylock A, Björkerud S, Brattsand R, Hansson GK, Hansson H-A, Bondjers G (1977) Endothelial structure in rabbits with moderate hypercholesterolemia. Acta Pathol Microbiol Scand [A] 85: 671-682

Chandler AB, Pope JJ (1975) Arterial thrombosis and atherogenesis, a survey of the frequency of incorporation of thrombi into atherosclerotic plaques. In: Hautvast JGAJ, Hermus RJJ, Van Der Haar F (eds) Blood and arterial wall in atherogenesis and arterial thrombosis. ET Brill, Leiden, p 110

Chien KR, Abrams J, Pfau RG, Farber JL (1977) Prevention by chlorpromazine of ischemic liver cell death. Am J Pathol 88: 539-558

Clark JM, Glagov S (1976) Luminal surface of distended arteries by scanning electron microscopy: elimination of contiguational and technical artefacts. Brit J Exp Pathol 57: 129-135

Davies PF, Bowyer DE (1975) Scanning electron microscopy: arterial endothelial integrity after fixation at physiological pressure. Atherosclerosis 21: 463-470

deDuve C (1964) Lysosomes and cell injury. In: Thomas L, Uhr JW, Grant L (eds) Injury, inflammation and immunity. Williams and Wilkins, Baltimore, p 283

Ericsson JLE, Brunk UT (1977) Alterations in lysosomal membranes as related to disease processes. In: Trump BF, Arstila AU (eds) Pathobiology of cell membrane, vol I. Academic Press, New York, p 217

Farber JL, Chien KR, Mittnacht S (1981) The pathogenesis of irreversible cell injury in ischemia. Am J Pathol 102: 271-281
Florentin RA, Nam SC, Lee KT, Thomas WA (1969) Increased ^{3}H-thymidine incorporation into endothelial cells of swine fed cholesterol for three days. Exp Mol Pathol 10: 250-255
Fowler S, Shio H, Haley NJ (1979) Characterization of lipid-laden aortic cells from cholesterol-fed rabbits. IV. Investigation of macrophage-like properties of aortic cell populations. Lab Invest 41: 372-378
French JE (1966) Atherosclerosis in relation to the structure and function of arterial intima, with special reference to the endothelium. Int Rev Exp Pathol 5: 253-353
Fry DL (1968) Acute vascular endothelial changes associated with increased blood velocity gradients. Circ Res 22: 165-197
Gaynor E (1971) Increased mitotic activity in rabbit endothelium after endotoxin. An autoradiographic study. Lab Invest 24: 318-320
Gerrity RG, Naito HK, Richardson M, Schwartz CJ (1979) Dietary induced atherogenesis in swine: morphology of the intima in pre-lesion stages. Am J Pathol 95: 775-786
Goldberg ID, Stemerman MB, Handin RI (1980) Vascular permeation of platelet factor IV after endothelial injury. Science 209: 610-612
Hansson GK, Bondjers G (1980) Endothelial proliferation and atherogenesis in rabbits with moderate hypercholesterolemia. Artery 7: 316-329
Hansson GK, Bondjers G, Nilsson L-Å (1979) Plasma protein accumulation in injured endothelial cells. Immunofluorescent localization of IgG and fibrinogen in the rabbit aortic endothelium. Exp Mol Pathol 30: 12-26
Hansson GK, Bondjers G, Bylock A, Hjalmarsson L (1980) Ultrastructural studies on the localization of IgG in the aortic endothelium and subendothelial intima of atherosclerotic and non-atherosclerotic rabbits. Exp Mol Pathol 33: 302-315
Hansson GK, Björnheden T, Bylock A, Bondjers G (1981) Fc-dependent binding of monocytes to areas with endothelial injury in the rabbit aorta. Exp Mol Pathol 34: 264-280
Hansson GK, Schwartz SM (1982a) Evidence for cell death in the aortic endothelium. Fed Proc (in press)
Hansson GK, Bondjers G, Schwartz SM (1982) Focal cell death in the arterial endothelium (submitted)
Hansson GK, Schwartz SM (1982b) Endothelial dysfunction without cell loss. In: Cryer A (ed) Biochemical interactions at the endothelium. Elsevier/North Holland Biomedical Press, London (in press)
Hawkins HK, Ericsson JLE, Biberfeld P, Trump BF (1972) Lysosome and phagosome stability in lethal cell injury. Am J Pathol 68: 255-288
Hugod C, Hawkins LH , Kjeldsen K, Thomsen HK, Astrup P (1978) Effect of carbon monoxide exposure on aortic and coronary intimal morphology in the rabbit - a reevaluation. Atherosclerosis 30: 333-342
Jennings RB, Ganote CE, Reimer K (1975) Ischemic tissue injury. Am J Pathol 81: 179-198
Kane AB, Stanton RP, Rayond EG, Dobson ME, Knafelc ME, Farber JL (1980) Dissociation of intracellular lysosomal rupture from the cell death due to silica. J Cell Biol 87: 643-651
Kunz J, Schreiter B, Schubert B, Voss K, Krieg K (1978) Experimentelle Untersuchungen über die Regeneration der Aortenendothelzellen. Automatische und visuelle Auswertung von Autoradiogrammen. Acta Histochem 61: 53-63
Laiho KU, Trump BF (1977) The relationship between cell viability and changes in mitochondrial ultrastructure, cellular ATP, ion and water content following injury of Ehrlich ascites tumor cells. Virchows Arch Cell Pathol 15: 264-277
Linder E (1981) Binding of Clq and complement activation by vascular endothelium. J Immunol 126: 648-658
Mittnacht S Jr, Sherman SC, Farber JL (1979) Reversal of ischemic motichondrial dysfunction. J Biol Chem 254: 9871-9878
Mustard JF, Packham MA, Kinlough-Rathbone RL (1981) Platelets, atherosclerosis and clinical complications. In: Moore S (ed) Vascular injury and atherosclerosis. Dekker, New York, p 79

Reidy MA, Bowyer DE (1977) Scanning electron microscopy. Morphology of aortic endothelium following injury by endotoxin and during subsequent repair. Atherosclerosis 26: 319-328
Reidy MA, Schwartz SM (1980) Developments in the study of endothelial cells by scanning electron microscopy. Artery 8: 236-243
Reidy MA, Schwartz SM (1981) Endothelial regeneration. III. Time course of intimal changes after small defined injury of rat aortic endothelium. Lab Invest 44: 301-308
Reidy MA, Schwartz SM (1982) Endothelial regeneration. IV. The effect of endotoxin on arterial endothelium. Lab Invest (in press)
Ross R, Harker L (1976) Hyperlipidemia and atherosclerosis. Chronic hyperlipidemia initiates and maintains lesions by endothelial cell desquamation and lipid accumulation. Science 193: 1094-1100
Ross R, Schwartz SM (1982) Endothelial interactions. In: Fisher AB, Fishman AP (eds) Handbook of physiology, vol IV. The Americal Physiological Society, Bethesda (in press)
Schwartz SM (1978) Selection and characterization of bovine aortic endothelial cells. In Vitro 14: 966-980
Schwartz SM, Benditt EP (1973) Cell replication in the aortic endothelium: a new method for study of the problem. Lab Invest 28: 699-707
Schwartz SM, Benditt EP (1976) Clustering of replicating cells in aortic endothelium. Proc Nat Acad Sci 73: 651-653
Schwartz SM, Benditt EP (1977) Aortic endothelial cell replication. II. Effects of age and hypertension in the rat. Circ Res 41: 148-255
Schwartz SM, Gajdusek CM, Reidy MA, Selden SC, III, Haudenschild CC (1980) Maintenance of integrity in aortic endothelium. Fed Proc 39: 2618-2625
Schwartz SM, Stemerman MB, Benditt EP (1975) The aortic intima. II. Repair of the aortic lining after mechanical denudation. Am J Pathol 81: 15-52
Spangoli LG, Pietra GG, Villuschi S, Johns LW (1981) Morphometric analysis of gap junctions in regenerating arterial endothelium. Lab Invest 46: 139-148
Stemerman MB (1981) Effects of moderate hypercholesterolemia on rabbit endothelium. Arteriosclerosis 1: 25-32
Svendsen E (1979) Focal endothelial injury in rabbit aorta. Aggravation of injury by two days of cholesterol feeding. Acta Pathol Microbiol Scand [A] 87: 123-130
Thomas WA, Reiner JM, Florentin RA, Scott RF, Lee KT, Janakidevi K (1981) Population dynamics of arterial cells in atherogenesis. In: Moore S (ed) Vascular injury and atherosclerosis. Dekker, New York, p 111
Trump BF, Arstila AU (1975) Cell membrane and disease processes. In: Trump BF, Arstila AU (eds) Pathobiology of cell membranes, vol I. Academic Press, New York, p 1
Velican C, Velican D (1979) Study on the onset of atherosclerotic lesions in human coronary arteries. Rev Roum Med-Med Int 17: 131-149
Weber G, Fabbrini P, Resi L (1974) Scanning and transmission electron microscopy observations on the surface lining of aortic intimal plaques in rabbits on a hypercholesterolic diet. Virchows Arch Pathol Anat 364: 325-331
Wright HP (1972) Mitosis patterns in aortic endothelium. Atherosclerosis 15: 93-100

Discussion

BORN: If your observations are transferable to the in vivo situation in, say a human aorta, the proposition of platelets adhering to the vessel wall, releasing their factor into the wall and getting it going on the smooth muscle, then becomes even more unlikely than I have thought for several years. Would you comment on that?

S. SCHWARTZ: First of all, my prejudice, and it is just a prejudice, is the same as yours. I do not think denudation is initiating lesion formation. The laboratory has now looked at several acres of endothelium, and they do not ever show anything that looks like denudation early. On the other hand there is absolutely no doubt from classical pathology, anyone who has ever looked at a plaque, that platelets do form thrombi on plaques. I think that as lesions progress, endothelium does break down, and thrombosis does occur. For adults, I mean all of us here, that may be more important than what happens when we are ten years old or twenty years old. I am very interested in why the endothelium might break down; but all I am saying for now is that in the initial stages the normal turnover and the normal cell death does not appear to de denudative.

WEINBAUM: Do you have any idea of the extent of permeability of the cell during the three days of its last throws?

S. SCHWARTZ: The fact that the IgG is in the cell means that the membrane must essentially be freely permeable. I would guess that in terms of your kind of modelling, the cell is one big hole. I would be surprised if it could contain anything. Permeability changes might not only occur through the dying cell, but also around the periphery. We do not know what happens to the junctions during that process.

DEWEY: The kinds of spiculations that we saw in these transient phenomenon were not associated with cell death. But once in a while a cell would go "wild" and you would see a great deal of internal motion. I think it is very interesting that, in the case where we apply the shear and then the cells accomodate, the motion will go on and then will decrease in time and pretty soon you get to this very static layer which looks just like your static conditions here. Hopefully within the next month we will have data in which we have an oscillating shear that cycles at 1 Hz from one dyne to eight dynes per square centimeter on a chronic basis. I think it is going to be very interesting to find out whether the cells see what is the equivalent of a change in shear because you are oscillating the shear on a continuous basis, or whether in fact they accomodate to the oscillations as if the time-mean value of shear was present with no oscillation. That could tell us a lot about the internal mechanics of all this.

S. SCHWARTZ: The video tape I showed you lasted for 24 hours, and if you draw on the screen an outline of the cell, it will not stay within the outline. There is some instability of the system. But if you draw three cells, those outlines move in the same direction. So it is something in the system that is moving and not the cells. I think it is important how long you leave cells down. The movie I showed you was of cells that had been down for two weeks at a stationary density before we began the experiment.

BLACKSHEAR: Have you speculated on the role of phagocytes in this process and what place would they intervene?

S. SCHWARTZ: When we put monocytes on the culture, they do a lot of interesting things, but they do not seem to be at home on these cells in culture. Beyond that I do not know.

FREUDENBERG: Have you any explanation for the mechanism of the migrating necrotic parts of endothelial cells on the surface you showed in your fine pictures?

S. SCHWARTZ: These cells have been dead for at least three days by the time they are doing that, and I guess what fascinates me is where are the energy sources coming from. I will tell you the two throughts I have had. One is the entropy of the membranes themselves. If the composition of the membranes is changing, that might make for rather dramatic changes in the shape of the cell, and this might take time as such changes sometimes do. The second possibility for which there is some data is a change in the state of the cyto-skeleton. That would also not require energy.

SCHMID-SCHÖNBEIN: Gert Schmid-Schönbein, my brother, has come up with a very simple idea as to how cells migrate and get fixed as a function of the substrate. One could very easily explain along the lines of his idea that these cells detach themselves irregularly from all sides. This is something that has to do with the cells having changed visco-elastic properties and that their cytoplasmic viscosity really decreases during times of activation. If you have a force which acts uni-directionally, the change of reattachment parallel to that force is much higher than that to the other sides. So that if the cell detaches itself from the substrate, it becomes reattached by the shear forces, and you could very easily explain that these cells are neatly oriented. This then ends itself due to mechanical forces once it has really reached the point where the membrane now limits the uprise.

S. SCHWARTZ: That is fine, and it may explain how thing move along on surfaces, but there are some little guys inside that cell who are trying to punch their way out.

Cellular Participation in Atherogenesis: Selected Facets of Endothelium, Smooth Muscle, and the Peripheral Blood Monocyte

C. J. Schwartz, E. A. Sprague, S. R. Fowler, and J. L. Kelley

A. Endothelium and the Evan's Blue Model

An outside feature of continuous arterial endothelium is its barrier function, whereby the influx of plasma constituents into the arterial intima is regulated. Of particular importance is the observation that arterial endothelial permeability is topographically non-homogeneous, and that areas of spontaneously-differing permeability to plasma macromolecules can be consistently identified in a number of species including pig, dog, and baboon (McGill et al. 1975; Packham et al. 1967; Bell et al. 1974a; Bell et al. 1974b; Sprague et al. 1980). Such areas of enhanced in vivo endothelial permeability, identified by their uptake of the protein-binding azo dye Evans Blue, are considered to be the sites of predilection for subsequent atheromatous development (McGill et al. 1957; Fry 1973; Schwartz et al. 1978). Because of the structural, hemodynamic, and functional implications of the Evans Blue model in atherogenesis, some of the salient features of these pre-lesion areas, characterized by the uptake of Evans Blue, are summarized in Table 1. Clearly, the aortic uptake of both ^{131}I-albumin and ^{131}I-fibrinogen at 2 hours is some 60% greater in blue than in white areas (Bell et al. 1974 a,b). Although the aortic uptake of ^{3}H-unesterified cholesterol is likewise greater in blue than in white areas, this difference in all likelihood represents a non-energy dependent physico-chemical cholesterol exchange, rather than a measure of lipoprotein influx alone (Somer and Schwartz 1971, 1972). However, as seen in Table 1, net intimal cholesterol accumulation in cholesterol-fat fed animals is some two-fold greater in blue than in white areas (Day et al. 1974; Bell et al. 1972), consistent with a greater influx and/or retention of cholesterol-rich low density lipoproteins.

Table 1

Aortic Sampling Site	^{131}I-Albumin Uptake (Relative %)	^{131}I-Fibrinogen Uptake (Relative %)	^{3}H-Unesterified Cholesterol Uptake (Relative %)	Intimal Cholesterol Accumulation (Relative %)	Endothelial Cell Turnover (Relative %)	Endothelial Cell Injury/ Death (%)
Thoracic White	100	100	100	100	100	0.71
Thoracic Blue	166.4	164.3	146.6	200	137	2.91

Fluid Dynamics as a Localizing Factor for Atherosclerosis
Editors: G. Schettler et al.

It is of particular interest that these pre-lesion or blue areas, which are characterized by an enhanced endothelial permeability to plasma proteins, also exhibit a significantly greater endothelial cell turnover as measured by ^{3}H-thymidine nuclear labelling (Caplan and Schwartz 1973) (Table 1). This observation we interpret as reflecting greater focal fluid mechanically-induced stress and injury in these areas. Whether such injury is the direct result of high shear stress, stretch, vibration, or pressure variations, or alternatively is the outcome of constituents of platelets or leukocytes released at or near the endothelium in areas of disturbed blood flow has yet to be determined. Consonant with the greater endothelial turnover in blue areas, together with the greater phospholipid synthesis which presumably reflects enhanced membrane synthesis and repair (Somer et al. 1974), more direct evidence of an enhanced endothelial cellular injury in these pre-lesion areas has been obtained using silver nitrate-stained Häutchen preparations. Silver impregnated cells, presumably reflecting membrane leakage, were found to be some four-fold more frequent in pre-lesion areas (Table 1) indicating that these sites are indeed sites of enhanced endothelial injury and/or cell death (Caplan et al. 1974). Whether such a frequency of cell injury or death (2.91% in blue relative to 0.71% in white areas) can account wholly or in part for the permeability differences as has been suggested (Weinbaum et al. 1980), need further clarification.

Whatever the nature of the intermediate mechanisms involved, it is reasonably clear that arterial endothelium exhibits striking focal differences in permeability, cell turnover, and injury or death, differences which we consider to represent focal differences in fluid mechanical stress, and which in all likelihood play a critical role in the initiation and augmentation of atherogenesis.

B. The Monocyte-Macrophage

In addition to the endothelium, the peripheral blood monocyte is another important cellular participant in atherogenesis. There are a number of mechanisms through which these cells may contribute to lesion development: first, because of their phagocytic capacity, as progenitors of macrophages, they may be an important source of intimal foam cells; second, in modulating the arterial immune system; third, in the uptake and metabolism of lipoproteines; fourth, the secretion of a number of hydrolytic and proteolytic enzymes; fifth, the synthesis and release of a mitogen for intimal smooth muscle cells; and finally, the secretion and release of lipoprotein lipase. Although monocytes are recruited to the normal arterial intima in relatively small numbers, recruitment is enhanced in hyperlipidemic animals (Schwartz et al. 1978; Gerrity et al. 1979; Schwartz et al. 1980). Whether the enhanced monocyte recruitment reflects an increased responsiveness to chemoattractants, or alternatively an influence of hyperlipidemia on the production of chemoattractant activity has yet to be determined. Recent findings (Jauchem et al. 1982) indicate that intimal smooth muscle cells may provide one source of arterial wall chemoattractant activity. Cultured aortic medial smooth muscle cells, but not aortic endothelial cells, produce chemoattractant activity for isolated blood monocytes (Fig. 1). This activity is relatively heat stable, of low molecular weight, and is not abolished by anti-sera to C_5 or C_3. The nature of this SMC-generated chemoattractant is currently under investigation.

Several aspects of arterial monocyte recruitment deserve brief comment. It is of some interest that recruitment appears to occur prefer-

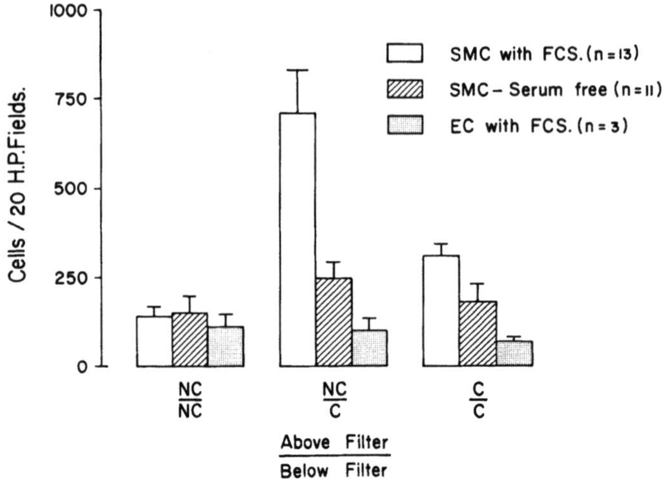

Fig. 1. Influence of SMC-conditioned media with and without serum (FCS), and EC-conditioned media containing serum on mononuclear cell migration. The data represent the means and standard errors of the means. The left three columns represent control or background migration values for each of the 3 categories of conditioned media respectively, where NC was placed both above and below the filters. *SMC*, aortic medial smooth muscle cells; *EC*, aortic endothelial cells; *FCS*, fetal calf serum; *NC*, non-conditioned medium; *C*, conditioned medium. (From Jauchem et al., Exp. Molec. Pathol., 1982, in press)

entially in prelesion or Evans Blue areas in the pig aorta (Gerrity et al. 1979), suggesting a possible hemodynamic influence. Monocytes in the process of migration through the endothelium do not exhibit a uropod which might give some indication of direction of movement, as is frequent with lymphocytes, and the disposition of the endothelial margins surrounding migratory cells appears to be an unreliable directional guide. Ultrastructurally, we have on occasions observed an endothelial discontiuity at the sites of cellular migration, an observation which suggests that monocyte migration may interfere with the normal endothelial permeability barrier function. Whatever the mechanisms for monocyte attachment to the endothelium, and their subsequent migration to the arterial intima, current studies clearly indicate that the monocyte-derived macrophages of experimental granulomas avidly accumulate lipid in hyperlipidemic animals, consistent with their role in the genesis of some intimal foam cells. In this context it is important to recognize that elicited macrophages exhibit both receptor-mediated binding and internalization of low density lipoproteins (LDL) (Fig. 2), as well as a low-affinity non-receptor mediated pathway (Fig. 3), providing at least a partial explanation for the evolution of the lipid-rich foam cell.

C. Endocytosis in Cultured Vascular Cells

Endocytosis, of which there are three types, namely receptor-mediated, adsorptive, and fluid phase, is important not only in the uptake of molecules by cells, but may also be critical in transcellular transport or permeability. When cells are exposed briefly to a cationic ligand, such as cationized ferritin (CF), anionic surface binding sites are visibly redistributed (Danon and Skutelsky 1976; Pelikan et al.

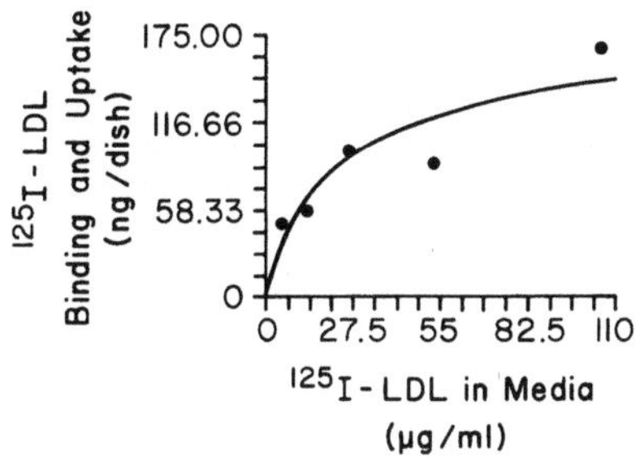

Fig. 2. Receptor-mediated binding and uptake of ^{125}I-LDL by thioglycollate elicited rabbit peritoneal macrophages. Macrophages were incubated for 4 hours at 37°C

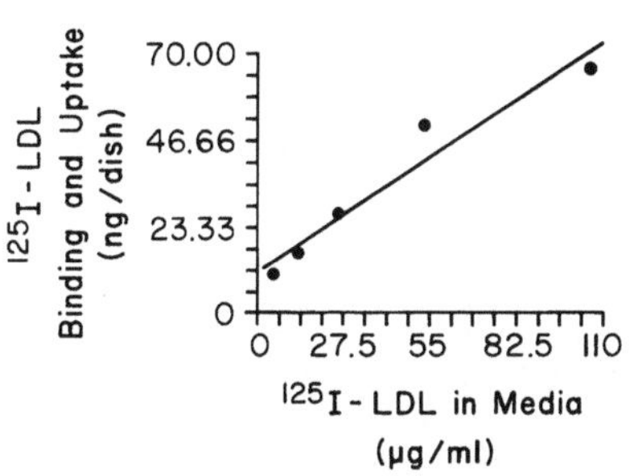

Fig. 3. Non-receptor-mediated binding and uptake of ^{125}I-LDL by thioglycollate elicited rabbit peritoneal macrophages. Macrophages were incubated for 4 hours at 37°C with increasing concentrations of labeled LDL and a 50-fold excess of unlabeled LDL in the incubation media

1979; Sprague et al. 1980), and an enhanced binding and internalization of molecules such as albumin and LDL is observed (Sprague and Schwartz 1981). Table 2 illustrates the striking influence of a 1-minute exposure to CF on albumin binding and internalization in cultured aortic medial smooth muscle cells (SMC). No effect of CF on ^{14}C-sucrose internalization was observed. An essentially similar phenomenon has been observed with cultured human umbilical vein endothelium (Fig. 4), where it can be seen that a 1-minute exposure to CF results in a significantly increased internalization of ^{125}I-albumin.

Table 2

		^{125}I-Albumin Binding (pMol/10^6 cells)		^{125}I-Albumin Internalization (pMol/10^6 cells)	
		Time (minutes)			
Treatment	Temperature	3	60	3	60
Cationized Ferritin (n = 5)	37°C	13.78	27.13	9.58	33.13
Native Ferritin (n = 5)	37°C	5.75	7.00	2.51	2.51
Cationized Ferritin (n = 5)	4°C	5.97	9.43	1.43	5.77
Native Ferritin (n = 5)	4°C	3.19	6.75	0.63	1.95

In other words, modification of cell surface anionic or negatively charged sites by a cationic ligand has a significant influence on adsorptive but not fluid phase (sucrose) endocytosis. This effect, seen both in cultured vascular endothelial and smooth muscle cells, conti-

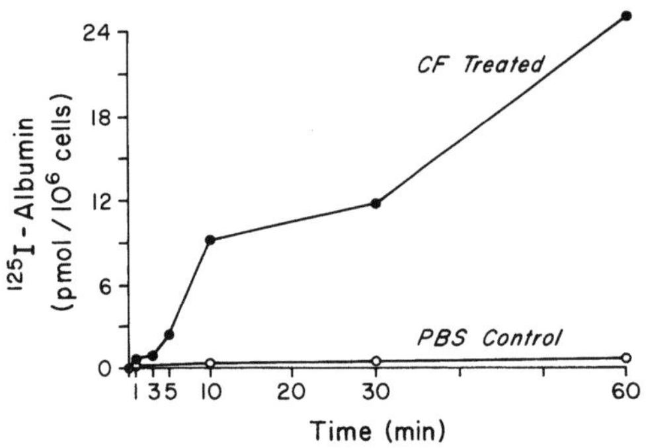

<u>Fig. 4.</u> Internalization of ^{125}I-albumin in confluent cultured human umbilical vein endothelial cells pretreated with 0.2 mg/ml cationized ferritin (CF) or PBS for 1 min at 37°C

nues over a period of 1 hour after removal of CF, although exposure to the CF was only for 1 minute, and is not associated with morphologic or enzymic evidence of cellular injury.

Recent data (Fig. 5) indicate that CF pretreatment has a similarly dramatic effect on ^{125}I-LDL internalization in cultured SMC at both 30 and 60 minutes. Whether this effect has been mediated through activation of high affinity LDL receptors or through an alternate pathway has yet to be clarified. The implications of cell surface charge in the regulation of cellular endocytosis are obvious. Whether surface charge, charge distribution, and endocytosis can be modified fluid-mechanically or by endogenous cationic compounds such as platelet derived growth factor released at sites of injury, is being actively examined.

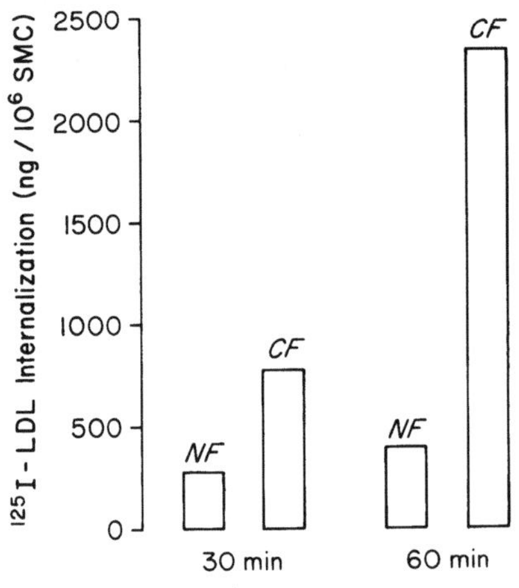

<u>Fig. 5.</u> Internalization of baboon ^{125}I-LDL in confluent cultured baboon aortic smooth muscle cells (SMC) pretreated with 0.2 mg/ml cationized (CF) or native (NF) ferritin for 1 min at 37°C

D. Summary

Areas of spontaneously differing endothelial permeability, cell turnover, and injury have been reviewed, with particular emphasis on initial events in atherogenesis, and the apparent close association of fluid mechanical factors to these events. Both the recruitment and the role of the blood monocyte in atherogenesis have been discussed, and the importance of cellular anionic surface charge distribution as a factor regulating the endocytosis of proteins including LDL by vascular cells has been presented.

Acknowledgements. This work was supported by NHLB1 grants HL-26890, HL-19362, HL-07446, HL-06075, and by the Texas Affiliate, American Heart Association, AHA-TX 81-24

References

1. Bell FP, Adamson IL, Schwartz CJ (1974a) Aortic endothelial permeability to albumin: focal and regional patterns of uptake and transmural distribution of ^{131}I-albumin in the young pig. Exp Molec Pathol 20: 57-68
2. Bell FP, Day AJ, Gent M, Schwartz CJ (1975) Differing patterns of cholesterol accumulation and ^{3}H-cholesterol influx in areas of the cholesterol-fed pig aorta identified by Evans Blue dye. Exp Molec Pathol 22: 366-375
3. Bell FP, Gallus AS, Schwartz CJ (1974b) Focal and regional patterns of uptake and the transmural distribution of ^{131}I-fibrinogen in the pig aorta in vivo. Exp Molec Pathol 20: 281-292
4. Caplan BA, Gerrity RG, Schwartz CJ (1974) Endothelial cell morphology in focal areas of in vivo Evans Blue uptake in the young pig aorta. I. Quantitative light microscopic findings. Exp Molec Pathol 21: 102-117
5. Caplan BA, Schwartz CJ (1973) Increased endothelial turnover in areas of in vivo Evans Blue uptake in the pig aorta. Atherosclerosis 17: 401-417
6. Danon D, Skutelsky E (1976) Endothelial surface charge and its possible relationship to thrombogenesis. An NY Acad Sci 275: 47-63
7. Day AJ, Bell FP, Schwartz CJ (1974) Lipid metabolism in focal areas of normal-fed and cholesterol-fed pig aortas. Exp Molec Pathol 21: 179-193
8. Fry DL (1973) Responses of the arterial wall to certain physical factors. In: Atherogenesis; initiating factors. Elsevier Publishing Co., Ciba Foundation Symposium 12: 93-125
9. Gerrity RG, Naito HK, Richardson M, Schwartz CJ (1979) Dietary-induced atherogenesis in swine. Morphology of the intima in prelesion stages. Amer J Pathol 95: 775-792
10. Jauchem JR, Lopez M, Sprague EA, Schwartz CJ (1982) Mononuclear chemoattractant activity from cultured arterial smooth muscle cells. Exp Molec Pathol (in press)
11. McGill HC, Heer JC, Holman RL (1957) Sites of vascular vulnerability in dogs demonstrated by Evans Blue. Arch Path 64: 303-311
12. Packham MA, Rowsell HC, Jorgensen L, Mustard JF (1967) Localized protein accumulation in the wall of the aorta. Exp Molec Pathol 7: 214-232
13. Pelikan P, Gimbrone MA, Cotran RS (1979) Distribution and movement of anionic cell surface sites in cultured human vascular endothelial cells. Atherosclerosis 32: 69-80
14. Schwartz CJ, Fine R, Kelley JL, Sprague EA, Jauchem JR (1980) Monocyte recruitment in the normal and atheromatous nonhuman primate aorta. Circulation 62, Suppl III: 96
15. Schwartz CJ, Gerrity RG, Lewis LJ (1978) Arterial endothelial structure and function with particular reference to permeability. In: Paoletti R, Gotto AM (eds) Atherosclerosis reviews. Raven Press, New York, 3: 109-124
16. Schwartz CJ, Somer JB, Gerrity RG (1978) Initial events in atherogenesis. In: McDonald L, Goodwin J, Resnekov L (eds) Very early recognition of coronary heart disease. Excerpta Medica, Amsterdam Oxford, pp 47-61
17. Somer JB, Bell FP, Schwartz CJ (1974) Focal differences in lipid metabolism of the young pig aorta. Synthesis from ^{14}C-1-acetale and ^{14}CU-glucose in vitro. Atherosclerosis 20: 11-21
18. Somer JB, Schwartz CJ (1971) Focal ^{3}H-cholesterol uptake in the pig aorta. Atherosclerosis 13: 293-304
19. Somer JB, Schwartz CJ (1972) Focal ^{3}H-cholesterol uptake in the pig aorta. Part 2. Distribution of ^{3}H-cholesterol across the aortic wall in areas of high and low uptake in vivo. Atherosclerosis 16: 377-388
20. Sprague EA, Frankel MA, Hagens MR, Wilhelm GB, Schwartz CJ (1980) Observations on vascular cell surface membrane charge distribution. In: Nerem RM, Guyton JR (eds) Hemodynamics and the arterial wall. University of Houston Press, 11-18

21. Sprague EA, Kelley JL, Schwartz CJ (1980) Focal patterns of ^{125}I-albumin uptake in the baboon aorta in vivo. Unpublished observations
22. Sprague EA, Schwartz CJ (1981) Influence of membrane surface charge distribution on ^{125}I-albumin binding and internalization in cultured arterial smooth muscle. Fed Proc 40: 327
23. Weinbaum S, Arminski L, Pfeffer R, Chien S (1980) Theoretical models for endothelial junction formation and vesicular transport. In: Nerem RM, Guyton JR (eds) Hemodynamics and the arterial wall. University of Houston Press, 31-33

Discussion

BORN: At my suggestion, Helen Payling Wright measured the turnover of altered endothelial cells and showed differences in turnover rates between bifurcations and elsewhere. We connected with Dr. Stehbens' observation of gaps between the dividing, but not between resting cells, and suggested that these gaps might account for an increase in lipoprotein efflux from the plasma. As Dr. Stehbens is here, I would like to know what he and you think about this? Also in view of Dr. Nerem's beautiful pictures, could one use such modern techniques to demonstrate differences in turnover rate around branches more precisely? We demonstrated this 15 years ago, but I should have thought one could do it much more elegantly nowadays.

C. SCHWARTZ: Let me just comment on the stigmata/stomata story. We did indeed find a greater number of stigmata and so-called stomata in the endothelium of the blue or pre-lesion sites relative to the white areas in the young normal pig. These were seen on Häutchen silver stain preparations and appeared either as little circular defects or as silver staining, small dots by definition. However, in our electron microscopic studies we have failed continuously to find any significant ultrastructural counterpart of the stigmata and stomata, and I do not honestly know whether these do indeed occur. The biological significance of these is something that I personally feel quite uncertain about. I certainly would like to hear Prof. Stehbens' comments on these because he was one of the people with the group in Oxford many years ago who became very interested in the morphology of endothelium.

STEHBENS: With regard to stomata and stigmata, the evidence certainly suggests that there is a correlation between stigmata and stomata on the one hand and areas of endothelial separation on the other. But it does not necessarily mean that every stigma and stoma is an area of cell separation. It may well be that it is an area that is functionally or physiochemically different where the cells may possibly separate from time to time. But I agree with you; for example, in my ultrastructural studies of the renal arterial bifurcation I have not been able to demonstrate endothelial separation. I have no Häutchen preparations on them, but certainly you find them at the origin of the segmental branches in the aorta. With regard to the shape of endothelial cells around the orifices of segmental branches in the aorta, there is increased turnover of endothelial cells. When endothelial cells multiply, the daughter cells will not be elongated in shape, but will be smaller and polyhedral in shape. Therefore it may well be that the polyhedral cells around the orifices of branches may be partially at least due to their greater turnover.

C. SCHWARTZ: It is quite obvious that in the blue areas in our studies there were a significantly greater number of more polyhedral rather than elongated and highly oriented cells which may be consistent with that idea. I guess that we should really go back and do these things again as suggested by Professor Born so as to have a much more precise look at sites of replication of cells in terms of vascular geometry.

S. SCHWARTZ: As far as the nature of stomata, using a high resolution technique, there appears to us to be no such thing as a discontinuity in the endothelium. The stigmata apparently represent areas where one endothelial cell overlaps on another. Thus, with the silver stain, it appears as a complete circle. I have no idea what that means functionally, although my guess would be that it means the surface is highly interfolded in that area. As far as stigmata, I do not think there is any morphologic equivalent of that by high resolution surface microscopy. There certainly are not any endothelial folds above a few hundred nanometers. The other thing I would like to comment on is turnover studies. I think it is very dangerous to equate turnover to injury. The endothelium does have focal areas of high turnover, but this may be quite normal as it is in gut. The gut has a very high level of turnover, yet maintains a total electrical barrier. I would also point out that a lot of the turnover studies, in fact the bulk of them, have been done in growing animals. It is impossible when one is seeing a replicating cell in a non-steady state system to be certain that we are looking at turnover.

C. SCHWARTZ: I would just like to respond very briefly indeed and say that I totally agree with your comments on stigmata and stomata. The interpretation for cell overlaps appears to be a very good one, and in terms of turnover, I think again it is very important not to equate turnover with cell injury. I think one needs to have concurrent information on cell injury or cell death in order to be able to interpret that data.

WEINBAUM: The experiments that we do in the Shu Chien's laboratory also confirm what Steve Schwartz just mentioned; however, there is another part of it which has not been brought up at all yet. This is that there are more subtle changes in the endothelial cells themselves; for example, under hypertension two key ones are the size of the particles that will enter vesicles and also the dimensions of the intercellular cleft, which until just recently we did not believe changed. We also know that the permeability of the cleft is altered by pressure in those experiments done just very recently. What I wanted to ask you was whether you have seen any difference in the filling of plasmalemma vesicles? Also, I was confused about how you use the word endocytosis.

C. SCHWARTZ: We use the term endocytosis in terms of three main categories. One is the so-called adsorptive endocytosis and we operational use this in terms of albumin binding and internalization in cells. The term receptor-mediated endocytosis in a very operationally way is defined, as an example, in terms of the binding and internalization of low density lipoprotein by cells, while fluid phase or bulk phase endocytosis is indeed the uptake of carbon-14 labelled sucrose. The studies that we have done are not marker studies; they are radio-isotope studies for binding and internalization. Until we get a little further with our ferritin-conjugated lipoproteins and our gold conjugates and other techniques, we will not be able to say just precisely whether these vesicles are being targeted, where they are, the numbers, and so on. We are not able to do that at this point.

Evidence that High Flow Velocity and Endothelial Disruption Are Not the Principal Factors in Experimental Plaque Localization

S. Glagov, C. K. Zarins, K. E. Taylor, R. A. Bomberger, and D. P. Giddens

Introduction

Evidence has been presented that experimental lesions tend to occur preferentially at sites presumed to be associated with elevated flow rates [1,2] and some investigators have contended that endothelial cell distortions and disruptions may occur at such sites as a result of high shear stress [3]. In an effort to detect morphologic evidences that endothelial injury is associated with lesion formation at sites of high shear, we have studied the distribution and evolution of diet induced plaques in rabbits and monkeys by means of scanning and transmission electron microscopy. We have evaluated the surface features of early lesions, as well as the localization of lesions and endothelial changes about branch ostia and in relation to experimentally produced aortic coarctions. We have also examined the effects of mechanically induced endothelial sloughing on lesion localization in cholesterol fed rabbits. Our findings indicate that high laminar flow rates tend to protect against lesion formation, that endothelial changes which have been identified at branch points may be artifacts of specimen preparation, and that endothelial disruption is more likely to be a consequence than a precursor of lesion induction.

Methods

Lesions were induced by feeding monkeys and rabbits standard atherogenic diets containing 2% cholesterol for periods up to 6 months. At the time of sacrifice, animals were anesthetized and their aortas cannulated. Fixation was carried out under conditions of controlled and monitored pressure distention by methods described in detail elsewhere [4]. Sections were prepared for light microscopy, as well as for scanning and transmission electron microscopy with minimal handling. Contact with the endothelial surface was completely avoided at sites of examination. *For the study of the aortic branch ostia,* samples were taken of the aortic ostia of intercostal and mesenteric arteries. Some were taken from aortas opened by 2 parallel lateral longitudinal incisions designed to preserve ostial configuration throughout processing, while others were opened by a single longitudinal cut and pinned flat [5]. *Coarctations* were produced in the descending thoracic aorta of monkeys either by suture plication or banding [6] and flow velocities were measured using pulsed Doppler ultrasound. *Endothelial sloughing* was accomplished by placing a 50% constricting ligature for 15 minutes about the aortas of the rabbits which had previously been maintained on atherogenic diets for 1 month or 3 months [7].

Results

Aortic Ostia. In pressure-fixed aortas opened by bivalving the endothelium was continuous and of normal surface configuration. Endothelial

Fluid Dynamics as a Localizing Factor for Atherosclerosis
Editors: G. Schettler et al.

cells showed no ultrastructural evidence of distortion or injury at flow dividers [5]. In contrast, we could regularly demonstrate fusiform distortions of endothelial cells on specimens pinned flat. These changes were indistinguishable from the "spindle cells" shown by others in pinned specimens [3] and offered as evidence of endothelial injury by high shear stress at the flow divider rim. Early foam cell lesions were not located on the rim of flow dividers where shear stress is high, but on the proximal and lateral inlet slopes of the ostium and in a triangular zone beyond the flow divider where shear stresses are lower [8]. Similar plaque localization at zones of relatively low flow rate and sparing where flow rates are high have been noted at the human carotid bifurcation [9].

Coarctation. Measurements of flow velocity at the time of coractation and before sacrifice revealed that flow velocity was markedly elevated within the coarctation channel compared to regions immediately proximal and distal to the constriction. Lesions did not form in the stenotic channels, nor was there any morphologic evidence of endothelial cell distortion or disruption within the channels six months after the constricting procedure [6]. If endothelial injury occurred acutely at the time of coarctation and/or if endothelial permeability was modified within the coarctation channel before or during the hyperlipidemic period, there was no lasting atherogenic effect. Lesions were abundant proximal and distal to the stenosis.

Endothelial Sloughing. The ligation injury resulted in endothelial sloughing and underlying necrosis of the media, but an intact endothelial surface was restored within 4 days. Lesions did not form preferentially at sites of injury and did not occur at all when the underlying media was markedly disrupted. Ligation injuries through lesions in atherosclerotic rabbits resulted in sloughing of the lesions but not in preferential recurrence of lesions at injury sites. Other investigators using injury models have noted that lesions from where endothelium is preserved or regenerated [10,11].

Lesion Evolution. Lesions consisting of a single layer of foam cells were covered by intact endothelium. Although the surfaces of most of the thicker lesions were intact, endothelial attenuation and focal discontinuities of the endothelial lining were seen, but only over lesions which were several foam cells thick [12]. Lesion free areas showed no evidence of endothelial deformation or disruption.

Conclusions

1. Focal endothelial disruption does not necessarily result in plaque formation in hyperlipidemic animals. On the contrary, the evidence suggest that an intact endothelial surface, and perhaps an intact media as well, are necessary for plaque formation. Furthermore, endothelial disruption over a plaque appears to be a consequence or complication of plaque growth.
2. High laminar flow rates do not normally result in endothelial distortion or disruption and do not appear to predispose to plaque formation. On the contrary, the evidence suggests that high flow rates may protect against plaque induction.

Acknowledgements. This work was supported in part by grants 15082 Atherosclerosis-SCOR of the National Institutes of Health and by grant CME 7921551 of the National Science Foundation.

References

1. Fry DL (1973) Response of the arterial wall to certain physical factors. In: Atherogenesis initiating factors, a Ciba Foundation Symposium (vol 12). New series, Elsevier, Amsterdam
2. Cornhill JF, Roach MR (1976) A quantitative study of the localization of atherosclerotic lesions in the rabbit aorta. Atherosclerosis 23: 489-501
3. Reidy MA, Bowyer DE (1977) Scanning electron microscopy of arteries: The morphology of aortic endothelium in haemodynamically stressed areas associated with branches. Atherosclerosis 26: 181-194
4. Clark JM, Glagov S (1976) Luminal surface of distended arteries; eliminating configurational and technical artefacts. Br J Exp Path 57: 129-135
5. Zarins CK, Taylor KE, Bomberger RA, Glagov S (1980) Endothelial integrity at aortic ostial flow dividers. Scanning electron microscopy, part III, pp 249-254
6. Zarins CK, Bomberger RA, Glagov S (1981) Local effects of stenoses: Increased flow velocity inhibits atherogenesis. Circulation 64 (suppl II): 221-277
7. Bomberger RA, Zarins CK, Glagov S (1980) Medical injury and hyperlipidemia in the development of aneurysms or atherosclerotic plaques. Surg Forum 31: 338-340
8. Zarins CK, Taylor KE, Lundell MI, Glagov S (1978) Aortic ostial morphology and the localization of atherosclerotic lesions: preliminary observations. In: Nerem RM, Cornhill JF (eds) The role of fluid mechanics in atherogenesis. Proc. of Specialists Meeting, Columbus, p 5-1 to 5-4
9. Zarins CK, Giddens DP, Balasubramanian K, Sottiurai V, Mabon RF, Glagov S (1981) Carotid plaques localize in regions of low flow velocity and shear stress. Atherosclerosis 1: 358a
10. Chidi CC, Klein L, DePalma R (1979) Effect of regenerated endothelium on collagen content in the injured artery. S.G.&O. 148: 839-843
11. Minick CR, Stemerman MB, Insull W Jr (1979) Role of endothelium and hypercholesterolemia in intimal thickening and lipid accumulation. Am J Path 95: 131-158
12. Taylor KE, Glagov S, Lamberti J, Vesselinovitch D, Schaffner T (1978) Surface configuration of early atheromatous lesions in controlled-pressure, perfusion-fixed monkey aortas. In: Johari O (ed) Scanning electron microscopy, part III, pp 459-464

Discussion

WEINBAUM: I think it is important to realize that, when you say you have an intact endothelium in the healing process, the junctions during the regeneration of the endothelial layer are basically not complete yet. The endothelium is for all practical purposes perfectly porous, even though it seems to be intact. I do not know how long it takes for the junctional complexes to become restrictive again to the molecules, but certainly the model I talked about this morning would say that it is just like no endothelial cell is there in terms of permeability. For example, you saw that the uptake went up by a factor of 2 when only one cell was leaking around the junction in a field of 100. If you have every cell that is healing over a lesion having its junction leaking, the endothelium will behave just as if it was completely porous. It is just filled with a whole bunch of minute holes.

SCHWARTZ: In the morphologic studies, whatever one can tell from morphology, it takes months after the endothelium is regenerated for normal junctional complexes to reform. That is not quantitative data, it is objective morphologic data.

ANONYMOUS: What do you think was occurring on the surface of the lesion that has such a beautifully organized fibrous cap? What is going on there?

GLAGOV: We have looked there using very high resolution SEM, and these junctions from the surface look funny. They are more like shingles. The junction, at least on the surface, is simplified; but what it looks like when you cut through, I think somebody ought to do.

ANONYMOUS: I wanted to ask you about how to distinguish between the different processes that we are looking at. The initiation of the lesion, depending upon which camp you come from, is due to lipid insudation or cell proliferation. Those are not the same things, and while Weinbaum's work does perhaps explain how lipids get into the wall, it certainly does not explain how cells proliferate.

GLAGOV: One of my points is that there is no such thing as a lesion. There is a family of lesions with various properties, with various relations between endothelium and the underlying matrix, and the function of the fibrous cap is a real function on morphologic grounds.

Localization of Early Atherosclerotic Lesions in Carotid and Coronary Bifurcations in Humans – A Bifurcation of the High Shear Stress Hypothesis

A. D. Svindland and L. Walløe

Atherosclerotic lesions have a patchy distribution in the arterial wall. Branching regions and stenoses are favoured sites for the development of lesions (Rindfleisch 1867; Mitchell and Schwartz 1965). Different hypotheses have been proposed to explain the localization. According to Fry's shear stress hypothesis (1969, 1972) the early lesions are to be expected in regions with high blood velocities, while according to Caro's boundary layer diffusion hypothesis (1969, 1971) the early lesions should develop in regions with low velocities. Most animal experiments support Fry's hypothesis. However, histological differences between dietary induced vascular lesions in animals and human atherosclerosis have long been recognized. Moreover, until recently the detailed localization of the early atherosclerotic lesions in branching regions of arteries has not been studied in humans.

We have recently shown that the early lesions in the bifurcation of the brachiocephalic artery into the subclavian and the common carotid arteries have distinct localizations (Kjaernes et al. 1981). The early lesions start to develop on the outer walls of the bifurcation, while the inner walls downstream from the flow divider are left free as are also the lateral walls. Since these results clearly are at variance with results obtained in experimental animals, and since they also conflict with commonly accepted ideas about atherogenesis in humans (endothelial damage caused by high shear stress as the initial step, see for instance Fry 1972, and Ross and Glomset 1976), we wanted to make an additional study of another large arterial bifurcation in humans. The bifurcation of the common carotid artery was chosen (Meyer and Noll 1974; Solberg and Eggen 1971). In addition we wanted to examine a much smaller bifurcation; the bifurcation of the left main coronary artery was chosen because of its obvious clinical importance.

The right carotid artery with its two branches was collected from 37 autopsies of persons who had suffered a violent or sudden death from a non-vascular cause. This selected material included persons 18-49 years of age. The median age was 32 years, 10 were females and 27 males. The common carotid artery was cut near the brachiocephalic artery, and the internal and external carotid arteries were cut perpendicular to their axis, 3.5 and 2.5 cm, respectively, from their origin. The method used to remove and open the arteries as well as the techniques used for staining and making photographs and drawings of the sudanophilic plaques were identical with those reported previously (Kjaernes et al. 1981). The information contained in the 37 drawings was fed to an Altos computer through a Summographics Bit Pad One digitizer, scaled to fit a standardized picture of the bifurcation, and added together.

The distribution of the early atherosclerotic lesions in the carotid bifurcation reveals a distinct pattern. Different regions of the artery show wide differences in the degree of atherosclerotic involvement. In some regions none of the 37 arteries had any lesions, while in other

Fluid Dynamics as a Localizing Factor for Atherosclerosis
Editors: G. Schettler et al.

regions more than 60 per cent of the arteries were involved. The peak values are clearly located at the outer walls of the bifurcation, while the inner walls as well as the apex of the fluid divider are well protected regions. At the lateral walls only a few of the 37 arteries have atherosclerotic lesions. In the carotid sinus the peak values are found along its periphery. Distally, however, the sinus is remarkably less involved. Together, the peaks make a special shape, like the letter U. In the common carotid artery the atherosclerotic lesions are found along a longitudinal ridge.

The coronary arteries for the study were collected from 21 adult individuals whose primary cause of death was not cardiac or vascular disease. The hearts were obtained intact during the post mortem examination. One tissue cube was removed from each heart. The cube consisted of the main left coronary artery (MS) and about 2 cm of the left circumflex artery (CX) and the left anterior descending artery (LAD) in addition to the underlying heart wall. The tissue cubes were fixed in 10% formalin, and 1.5 mm thick consecutive and parallel slices were cut perpendicular to the main coronary artery. From each tissue cube seven of these slices were examined, one from the bifurcation itself, three proximally to the bifurcation, and three distally to the bifurcation. The seven slices were embedded in paraffin, cut and stained with different stains. Black and white photographs were taken of the sections stained with Verhoff's stain, all in the same enlargement, and prints made in a large format. Transparencies were used to draw lines corresponding to the inner border of the intima (the endothelium) and the border between intima and media, the internal elastic lamina (IEL). Atherosclerotic lesions in the intima were marked on the drawing. Throughout these procedures the orientation of the arteries was carefully controlled, and all the differently stained sections were used for microscopic examination of the border lines and of the lesions.

The microscopic appearance of an atherosclerotic lesion was usually a fibrous plaque with accumulations of smooth muscle cells, large amounts of connective tissue like a cap around a deeper pool of tissue debris, foam cells, and spaces from lipoid crystals. A few of the lesions were more complicated with small spots of calcification, further degenerative changes and vascularization. The most simple type of atherosclerotic lesion appeared as accumulations of foam cells, small spots of cell debris and cells of inflammation.

The drawing of the cross sections of the arteries were digitized by a Summographics Bit Pad One digitizer and fed to an Altos computer system. The cross sections of the digitized vessels were transformed to a circular shape. The information on the atherosclerotic lesions in the intima, and on the thickness of the intima, was stored in the computer for 360 points on the IEL. This information was then pooled or averaged for each segment.

Casts were made of the left main coronary bifurcation from 10 additional hearts. Hearts with about 5 cm of aorta attached were obtained intact during post mortem examination. The left coronary arteries were exanguinated and then filled with a silicone rubber compound. The casts were then dissected free from the arteries, and the curvature of the arteries and the angles about the bifurcation were measured. The median value of the angle MS-CX was 107 degrees, of CX-LAD 73, of LAD-MS 167, and the median radius of the LAD-curvature was 4 cm.

Our main finding is a distinct spatial pattern of the atherosclerotic lesions in the coronary bifurcation. Again the lesions are found at the outer walls of the bifurcation, while the inner walls downstream from the bifurcation are left free. The lesions are also predominantly located at the myocardial side of the vessels, that is, along the inner curvature of the curved vessels. In our sample more than 70% of the arteries had atherosclerotic lesions on the outer walls of the bifurcation, while only some 10% had lesions on the lateral wall at the epicardial side. These findings are in complete agreement with our earlier results obtained from the bifurcation of the brachiocephalic artery into the subclavian and common carotid arteries (Kjaernes et al. 1981) and also with the results obtained from the carotid bifurcation. The localizations of the lesions at the outer walls in the proximal part of the LAD and CX are also in broad outlines consistent with the results obtained by Fox et al. (1982). However, the coordinate system they used in measuring the location of the lesions, had a very low spatial resolution.

Our results are obviously not in accordance with Fry's high shear stress hypothesis. The flow pattern in a bifurcation under physiological pulsatile flow conditions is not known in detail. However, our own preliminary results from laser doppler measurements on casts from bifurcations and from finite element calculations (Wille 1982) indicate that there are ventilated flat bubbles of recirculating blood along the outer walls in a bifurcation. Most experimental studies on steady and pulsatile flow in curved tubes illustrate the flow pattern that occurs in the entry region of the curvature. However, recent work both in our group (Bertelsen and Thorsen 1982) and by others (Chandran and Yearwood 1981) shows the velocity distribution at various locations in a curved tube, and trapped vortical motions are observed along the inner wall of the tube in a straight portion of the tube downstream from the curvature.

References

Bertelsen AF, Thorsen LF (1982) An experimental investigation of oscillatory visceous flow in pipe bends. J Fluid Mech 118: 269-284

Caro CG, Fitz-Gerald JM, Schroter RC (1969) Arterial wall shear sand distribution of early atheroma in man. Nature 223: 1159-1161

Caro CG, Fitz-Gerald JM, Schroter RC (1971) Atheroma and arterial wall shear. Proc Roy Soc Lond B 117: 109-159

Chandran KB, Yearwood TL (1981) Experimental study of physiological pulsative flow in a curved tube. Fluid-mech 111: 59-85

Fry DL (1969) Certain chemorheologic considerations regarding the blood vascular interface with particular reference to coronary artery disease. Circulation 39 (suppl 4): 38-57

Fry DL (1972) Localizing factors in arteriosclerosis. In: Likoff W, Segal BL, Insull W, Moyer JH (eds) Atherosclerosis and coronary heart disease. Grune and Stratton (New York), pp 85-104

Kjaernes M, Svindland A, Walløe L, Wille SØ (1981) Localization of early atherosclerotic lesions in an arterial bifurcation in humans. Act path microbiol scand Sect A 89: 35-40

Meyer WW, Noll M (1974) Gross patterns of early lipid deposits in the carotid artery and their relation to performed arterial structures. Artery 1: 31-45

Mitchell JRA, Schwartz CJ (1965) Arterial disease. F.A. Davis co., Philadelphia

Rindfleisch E (1867-69) Lehrbuch der pathologischen Gewebelehre. Engelmann, Leipzig, p 163; or: A textbook of pathologic histology. Translated by Kloman WC and Miles FT, 1872. Lindsay and Blakiston, Philadelphia, p 211

Ross R, Glomset J (1976) The pathogenesis of atherosclerosis. N Engl J Med 295: 369-377

Solberg LA, Eggen DA (1971) Localization and sequence of development of atherosclerotic lesions in the carotid and vertebral arteries. Circulation 43: 711-724

Wille SØ (1982) Finite element stimulations of pulsatile blood flow patterns. In: Gallagher, Simon, Johnson, Gross (eds) Finite elements in biomechanics. Wiley, pp 39-60

Discussion

FRIEDMAN: Could you go over once again how you identified that there was a lesion at a particular site? In particular, did you use any basis other than sudanophilia to characterize a site in either of those first two bifurcations as a lesion?

WALLDØE: In those first two bifurcations the most important information was sudanophilic. We also used the original preparation, and afterwards we made from some of the sudanophilic lesions some not very good histological sections. The borderlines we had indicated always corresponded to the borderline between formed cells and ordinary intima as we could see it in these not too good preparations. In the coronary arteries we used a different set of histological techniques and many different staining techniques to identify what we call atherosclerotic lesions.

SEED: It is just a word of caution. In some of the material we looked at, which is very similar to yours, when we looked histologically at sections in which there was sudanophilia, if we cut frozen sections we could identify the lesion satisfactorily because we could see the fat. If we used standard histological preparation techniques on the same material, we could not find the lesion. The fat was gone and the underlying or surrounding structure did not contain sufficient characteristic changes for us to identify the site of the lesion. So I think that there is a real risk that you may miss the very earliest lesions if you do not use either macroscopic fat staining techniques or frozen sections to identify where the fat is.

WALLDØE: Now you are talking about the coronary arteries, and we are going to use a similar technique, in fact we are doing that now. I cannot give you the results, but we are doing the sudanophilic technique on the coronaries just now, and we will compare the results afterwards.

BORN: Does the distribution correspond to late lesions? It is very important in view of what presumedly happens later.

SEED: In our study where we had fatty and fibrous lesions represented, we plotted and I showed you the distribution of the fatty lesions. When we separately took specimens which had fibrous disease present and plotted the topography of the fibrous lesions, we got an almost exact superposition. So I think that as far as we were concerned, we are confident that the fatty lesion and the fibrous lesion were superimposable. We did not see any more advanced lesions than that.

Quantitative Arterial Wall Morphology

J. F. Cornhill

The understanding of the etiology of atherosclerosis requires quantitative methods to evaluate objectively the *extent* and *location* of gross arterial lesions. If such methods of lesion mapping were available, it would be possible to conduct sequential studies on the progression and regression of both spontaneous and experimental atherosclerosis. The need is particularly great when considering any possible role that hemodynamics may play in atherogenesis. The supposition that hemodynamic forces are important in the initiation and localization of atherosclerotic lesions originated from the fact that lesions do not occur randomly in the systemic vascular system but that certain sites and geometric configurations (e.g. bends and branches) have a significantly higher possibility of developing the disease than do others. However, many of the observations on which this premise is based are anecdotal or semi-quantitative at best. Quantitative morphometric techniques which determine the exact location of atherosclerotic lesions over a population are essential if we are to determine with any degree of certainty if hemodynamic forces play an important role in atherogenesis.

Traditionally, the extent of atherosclerosis has been assessed by visually grading arteries as free from disease, slightly diseased, moderately diseased, or severely diseased whilst the location of the disease was based on the subjective [1,2,3]. Mitchell and Schwartz [4] in their comprehensive study of necropsy material made a significant contribution to the quantization of atherosclerosis. In these studies, data was obtained by planimetry and point counting to determine the percentage involvement of the aortic, carotid, iliac and coronary arteries with atherosclerosis, the location of the lesions (i.e. proximal, middle or distal portions), and in some cases the length of lesions. In addition, the percentage involvement was subdivided into fatty streaks, raised fatty plaques, fibrous plaques and complicated lesions. Unfortunately, the International Atherosclerosis Project [5] did not adopt the methodology of Mitchell and Schwartz [4] but estimated the percentage involvement by a simple visual grading system, a method which Cranston et al [6] had shown to be inaccurate.

The studies of Mitchell et al.[4], Schwartz et al.[7] and McGill et al. [5] presented interesting information - although not strictly quantitative - concerning the localization of lesions. These observations have led several investigators to suggest that hemodynamic forces such as shear stresses [8,9], boundary layer separation [10], turbulence [4,11] and others may be of importance in atherogenesis. These postulates have initiated the demand for more detailed pathological information on the *exact* location of lesions with respect to bends and bifurcations. The previous studies [4,5,7] failed to determine if the lesions occurred on the inner or outer bend of a curving vessel, proximal or distal to a bifurcation etc. This type of detailed quantitative observation is essential to determine the importance of hemodynamics in atherogenesis.

Fluid Dynamics as a Localizing Factor for Atherosclerosis
Editors: G. Schettler et al.

Localization of Atherosclerotic Lesions at Branches

The anecdotal information on the localization of lesions at branches and the expected large spacial variations in wall shear stresses have led to several studies on the *precise* location of lesions at arterial branches. Although semi-quantitative studies were conducted on aortic lesions at branches in man [9] and dogs and pigs [12], it was not until the polar coordinate method of quantifying lesions at ostia was developed that truly quantitative data was available [13]. To use this technique, vessels are opened longitudinally, fixed and processed in a saturated sudanophilic stain to demonstrate lipid accumulation. The centroid of the ostia is placed at the polar coordinate origin and oriented so that 0° corresponds to a known anatomical or physiological landmark, e.g. blood flow direction (Fig. 1). The ostium is traced; the distance between the ostial lip and the edge of the lesion is measured at different radial angles; and the data are plotted on rectangular coordinates. The method is easily computerized using digitizers interfaced with computers or image analysis systems. The polar coordinate method enables investigators to quantitate specific ostial lesions and to produce mean composite images representative of entire populations.

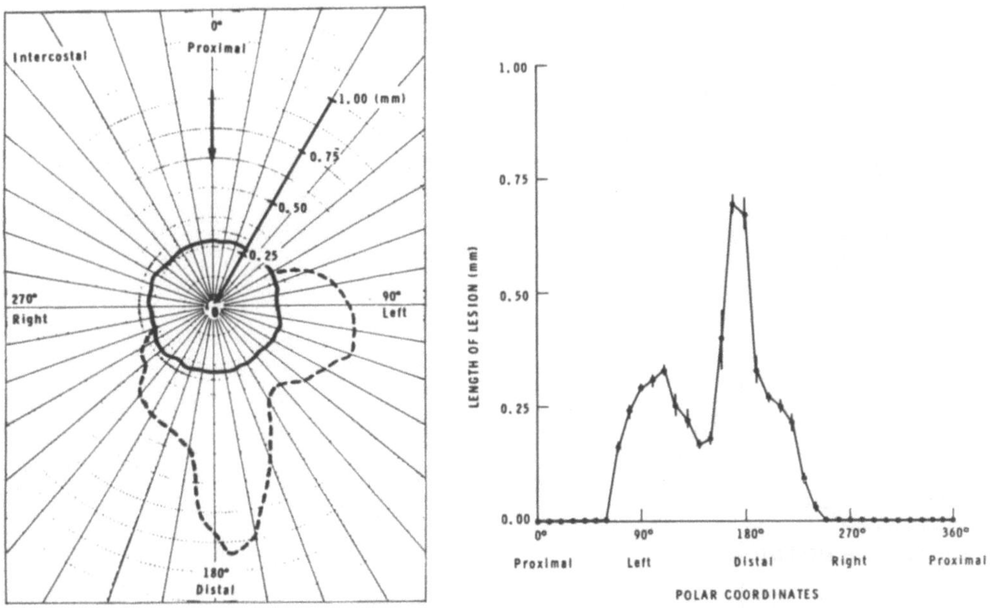

Fig. 1. Polar coordinate and rectangular coordinate graphs of an intercostal artery and associated lesion. In the polar coordinate drawing the sudanophilic lesion on the aortic wall (broken line) is distal to the intercostal ostia (solid line). In the rectangular coordinate graph the mean length and the standard error of the mean of the lesion on the aortic wall (i.e., differences between broken and solid lines) has been plotted against the polar coordinate angle; relative maxima occur at 100° and 170°. (From [14])

The polar coordinate technique was first used to quantify early sudanophilic lesions at branches from the aorta in the hypercholesterolemic rabbit [14]. The lesions on the aortic wall at the origin of the intercostal, superior and inferior mesenteric, coeliac and right and left renal arteries were found to occur in regions primarily distal to the ostia-areas where one would expect elevated wall shear stress as a result of a growth of a new boundary layer. Although the rectangular coordinate plots of the lesions at all of these ostia resemble, in general, the one in Fig. 1, some spatial skewing was observed particularly at branches which originated immediately distal to a major branch (e.g. left renal and inferior mesenteric ostia; Fig. 2). The exception to the uniform localization of lesions distal to branch points was the coronary lesion. This sudanophilic lesion completely surrounded the ostia with no predominate proximal or distal orientation (Fig. 3). It has been suggested that this distribution may result from unique hemodynamic conditions which may exist at the origin of the coronary arteries. These may include a) new boundary layers being formed at both the proximal and distal lips of the coronary ostia due to the re-attachment of the main stream to the aortic wall and the formation of vortices in the sinus of Valsalva; b) the bidirectional flow in the proximal aorta with reverse flow occurring, par-

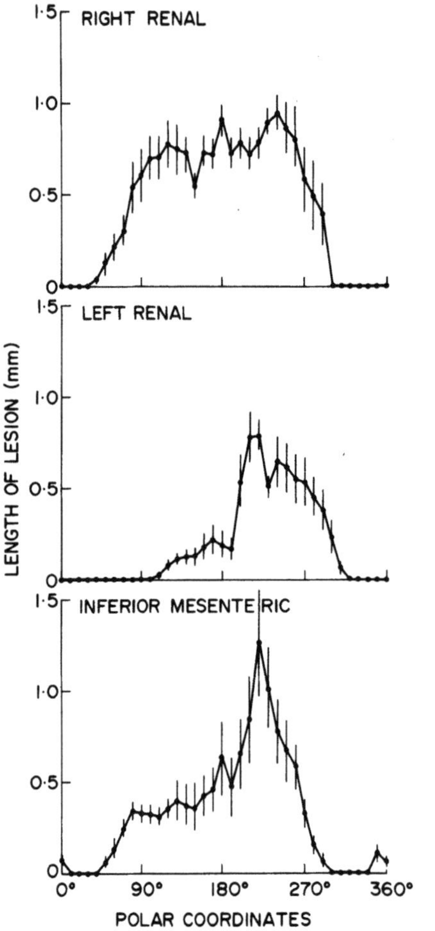

Fig. 2. Right renal, left renal and inferior mesenteric lesions. The lesions are distal to the ostia. The right renal lesion is symmetrical about 180° while the left renal and inferior mesenteric lesions are skewed to the right and have maxima at 220°. (From [14])

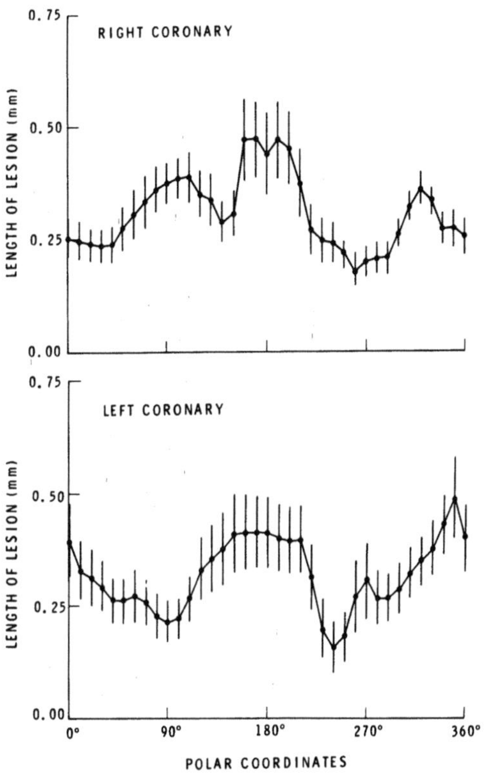

Fig. 3. Sudanophilic lesions completely surround the left coronary ostia. The left coronary lesion has relative maxima at 170° and 350°. Each point represents the mean ± SEM of 5 lesions. (From [14])

ticularly at the wall, as a result of the rapid deceleration of the fluid; c) the flow into the coronary arteries which is strangely biphasic with large accelerations at the beginning and end of diastole.

The progression of sudanophilic lesions in the hypercholesterolemic rabbit have been studied using the polar coordinate method [15]. In Fig. 4 the growth of lesions at the right renal ostia, which is typical of all ostial lesions in the descending thoracic and abdominal aorta is shown over a period of 6 months on a 0.25% cholesterol and 6% corn oil diet. The lesions first develop in the distal quadrant (i.e. 2 months), gradually grow into the right and left quadrants (i.e. 3 and 4 months) and finally grow into proximal quadrant (i.e. 6 months). Throughout this period of lateral spreading, the maximum length of the lesions occurred in the distal quadrant (i.e. ~ 180°). Sinzinger and co-workers observed a virtually identical pattern in the growth of sudanophilic lesions at the coeliac orifice of human fetuses and children [16].

In contrast, studies of the progression of spontaneous sudanophilic coeliac lesions on the aortic wall in White Carneau Pigeons, using the polar coordinated method, have shown that the lesions develop initially in the proximal quadrants (270° → 90°) with two relative maxima (~20° and ~ 320°). The lesions then grow proximally and radially with only limited circumferential spreading (Fig. 5). Little sudanophilic material was observed on the distal flow divider (90° → 270°) even in the oldest birds (6.5 years). The mode of spontaneous coeliac sudanophilic lesions development in the White Carneau Pigeon

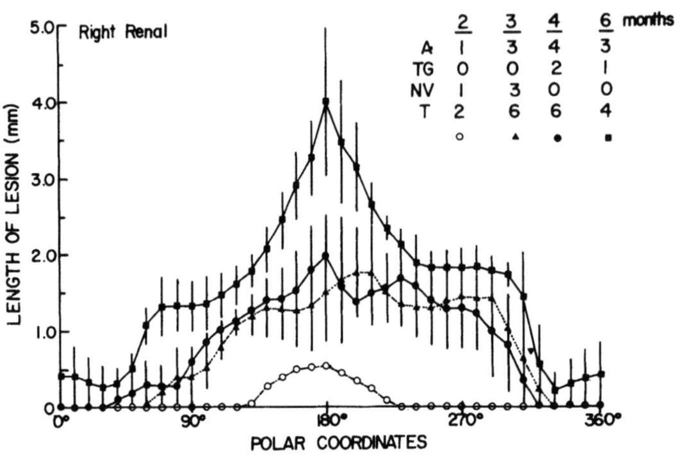

Fig. 4. The development of sudanophilic lesions at the right renal ostia on the low cholesterol diet. The lesions have a maximum at 180° (i.e. distal) with increasing proximal components as the period of atherogenic diet increases. Data points are the mean and standard error of the number of lesions shown in the upper right (A = analyzed; TG = too gross; NV = not visualized; T = Total). (From [15])

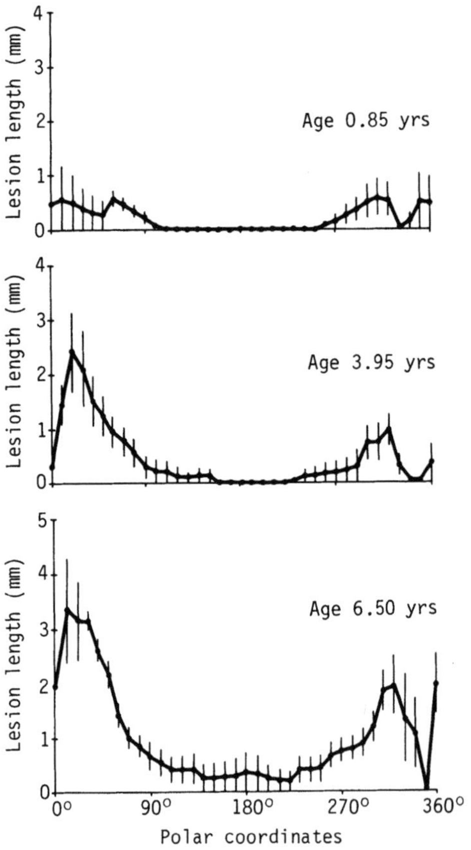

Fig. 5. Rectangular coordinate plot of length of celiac sudanophilic lesion as a function of polar coordinate angle. Zero degrees and 360° are proximal (i.e., upstream) to the celiac ostia and 180° is distal and then grow radially in the proximal region with some limited circumferential growth in the distal region with increasing age. (From [17])

is in sharp contrast to that observed both in spontaneous lesions in human infants and in experimental lesions in the hypercholesterolemic rabbit where lesions develop in the distal quadrants (90° → 270°); however, these data are consistent with the semi-quantitative data of Caro and co-workers [9] who found in humans of advanced age that coeliac atherosclerosis tended to be localized in the proximal region (i.e. 213° → 45°).

The studies presented above indicate that atherosclerotic lesions do occur at branches but that it is not possible to make an encompassing statement about the precise location of these lesions with respect to the ostia (i.e. proximal or distal). The paradox in these quantitative observations presents many questions to be answered regarding the role of hemodynamics in atherogenesis.

Localization of Non-Orificial Atherosclerotic Lesions

I. Probability Mapping

Although the efficacy of the polar coordinate method in the study of the localization, progression and regression of atherosclerotic lesions has been demonstrated clearly, the method is limited in that it can be applied only to lesions associated with ostia. As result, other methods have been developed that allow investigators to measure the surface characteristics of larger arterial segments, and from this data to construct probability maps of single or multiple types of atherosclerotic lesions [18]. Using this method, photographs of flattened arteries are analyzed for the presence or absence of specific types of lesions, that either are or are not stained, at approximately 1000 identical sites on all vessels. The probability of a lesion being present at each site is then calculated and a contour probability map constructed for an artery normalized for size. A computer-drawn topographic representation of the distribution of sudanophilic lesions in the opened human basilar artery is presented in Fig. 6. The extent of the sudanophilic lesions decreased as one proceeded distally from the origin of the basilar artery at the confluence of the vertebral bran-

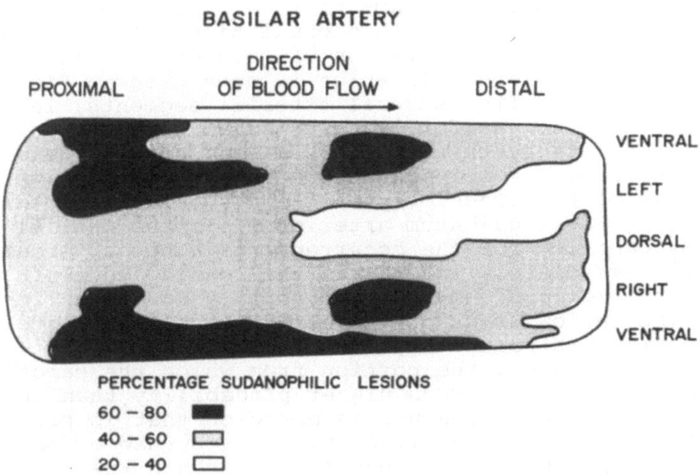

Fig. 6. Computer-drawn and smoothed topographic probability map of the average location of sudanophilic lesions in the human basilar artery of the 15 vessels studied. (From [18])

ches. Significantly more sudanophilic material was observed on the ventral (outer curvature) as opposed to the dorsal (inner curvature) surfaces. These data suggest that hemodynamic forces associated with confluent flow and curvature may be important in the localization of sudanophilic lesions in the proximal and ventral aspects of the human basilar artery.

This method of determining topographic probability from a standard vessel was produced by normalizing arterial length and width and is limited to short arteries or to arterial segments with relatively simple geometries. The technique cannot be applied to arteries with more complicated geometries such as the thoracic aorta, where simple normalization of vessel length and width potentially can obscure important spatial variations in lesion distribution. The method is also limited both by the subjective criteria used to determine the presence or absence of lesions at data fiducial points and by the time required to determine the nature of the lesions at thousands or tens of thousands of data fiducial points.

II. Image Analysis

The limitations of the methods presented above, namely the time consuming and subjective nature of data capture and the problems associated with normalization of vessel dimensions on the basis of length and width, have been overcome by developing an interactive image processing system for the analysis of the topographic distribution of atherosclerosis. A general technique has been developed in which a biological image may be divided into a number of triangular subsections with the vertices of triangles being anatomical landmarks such as ostia [19]. Using a simple linear transformation, the data may be transformed to a standard coordinate system to overcome local and global variations in shape and size among the individual specimens.

With the data transformed to a standard coordinate system, and with the use of image segmentation algorithms [20], it is possible to produce objective quantitative probability maps that represent the spatial variation of sudanophilic lesions within populations of arteries. A computerized, image processed representation of sudanophilic lesions in thoracic aortas of cholesterol-fed mini-pigs is presented in Fig. 7. This probability map, derived from several arterial segments, indicates that atherosclerosis is not uniformly distributed over the entire surface of the vessel, but rather, that there is strong spatial dependence with the highest probability for disease to be associated with large ostia such as those of the brachiocephalic and subclavian arteries, and in the region of the ligamentum arteriosum. In the ostial regions, the highest probability for the occurrence of atherosclerosis was observed in the area immediately lateral to the leading edge of flow dividers. The second region of highest probability was found immediately proximal to the inflow track of the ostia. Areas immediately distal to the larger ostia were relatively free from disease. The outer curvature of the aortic arch, the portion from which the carotid and subclavian arteries arise, had a much higher probability than the inner curvature. The distal inner curvature is a region that is particularly spared of atherosclerosis involvement. It is by use of these quantitative statistical maps that we may begin to approach the problem of correlating the spatial pattern of the disease with local chemistry and histology, and with hemodynamic influences.

Fig. 7. Topographic map of the probability of sudanophilic lesions in the thoracic aorta of cholesterol-fed mini-pigs, generated using image analysis and computer techniques. The origin of the aorta (i.e., aortic valve) is to the right; the descending thoracic aorta at T8 is to the left. Probability bands are in 15% intervals (i.e., 0-15%, 15-30%, etc.) with the lightest colors representing low probability and the darker shades representing the highest probabilities

Discussion

From the quantitative studies presented above, it appears that the localization of atherosclerotic lesions in the aorta has a significant spatial variation. The association of the disease with specific geometric entities such as branch points and major bends does exist. However, this correlation is far from overwhelming, and there does not appear to be a uniform localization of disease within these geometries. Regions of high and low probability of lesion occurrence may be found both proximal and distal to branches depending on the lesion type and species under study. In addition regions of high probability occur in regions not associated with bends or bifurcations, areas of expected hemodynamic disturbances.

Although hemodynamic forces may play a major role in the localization of some sudanophilic lesions, they do not at this time provide an encompassing explanation of the localization of the disease. The importance of other local biochemical and anatomical functions such as localized barriers to arterial wall transport, must also be investigated to account for the nonuniform distribution of the disease. Although, quoting from the title of the symposium, fluid dynamics may be an important localizing factor for atherosclerosis; it does not appear to be the localizing factor for atherosclerosis. Future work on the localization of the disease must take into account not only the other parameters that may be important in spatially varying mass transport but must also begin to investigate more carefully the mechanism whereby hemodynamics affect the arterial wall.

References

1. Gore I, Tejada C (1957) The quantitative appraisal of atherosclerosis. Am J Pathol 33: 875-885

2. Roberts JC, Moses C, Wilkins RH (1959) Autopsy studies in atherosclerosis. Circulation 20: 511-536
3. White NK, Edwards FE, Dry TJ (1950) The relationship of the degree of coronary atherosclerosis with age in men. Circulation 1: 645-654
4. Mitchell JRA, Schwartz CJ (1965) Arterial disease. Blackwell Scientific Publishing Ltd., Oxford, England
5. Guzman MA, McMahan CA, McGill HC Jr, Strong JP, Tejada C, Restrepo E, Eggan DA, Robertson WB, Solberg LA (1968) Selected methodologic aspects of the International Atherosclerosis Project. Lab Invest 18: 479-497
6. Cranston WI, Mitchell JRA, Russel RWR, Schwartz CJ (1964) Assessment of aortic disease. J Atheroscler Res 4: 29-39
7. Schwartz CJ, Ardlie NG, Carter R, Paterson JC (1967) Gross aortic sudanophilic and hemosiderin deposition. Arch Path 83: 325-332
8. Fry DL (1969) Certain chemorheologic considerations regarding the blood vascular interface with particular reference to coronary artery disease. Circulation 40, suppl 4: 38-57
9. Caro CG, Fitz-Gerald JM, Schroter RC (1971) Atheroma and arterial wall shear, observation, correlation and proposal of a shear dependent mass transfer mechanism for atherogenesis. Proc Roy Soc Lond B 177: 109-159
10. Fox JA, Hugh AE (1966) Localization of atheroma - a theory based on boundary layer separation. Brit Heart J 28: 388-399
11. Wesolowski SA, Fries CC, Sabini AM, Sawyer PN (1965) The significance of turbulence in hemic systems and in the distribution of the atherosclerotic lesion. Surgery 57: 155-162
12. Flaherty JT, Ferrans VJ, Pierce JE, Carew TE, Fry DL (1972) Localizing factors in experimental atherosclerosis. In: Likoff W et al. (eds) Atherosclerosis and coronary heart disease. Grune and Stratton, New York, NY, pp 40-83
13. Cornhill JF, Roach MR (1974) Quantitative method for the evaluation of atherosclerotic lesions. Atherosclerosis 20: 131-136
14. Cornhill JF, Roach MR (1976) A quantitative study of the localization of atherosclerotic lesions in the rabbit aorta. Atherosclerosis 23: 489-501
15. Roach MR, Cornhill JF, Fletcher J (1978) A quantitative study of the development of sudanophilic lesions in the aorta of rabbits fed a low-cholesterol diet for up to six months. Atherosclerosis 20: 259-264
16. Sinzinger H, Silberbauer K, Auerswald W (1979) Sudanophilic lesions of human aortic branches in infants: a quantitative evaluation using the polar coordinate mapping method. Atherosclerosis 33: 149-153
17. Cornhill JF, Levesque MJ, Nerem RM (1980) Quantitative study of the localization of sudanophilic coeliac lesions in the white carneau pigeon. Atherosclerosis 35: 103-110
18. Cornhill JF, Akins DE, Hutson MS, Chandler AB (1980) Localization of atherosclerotic lesions in the human basilar artery. Atherosclerosis 35: 70-79
19. Cornhill JF, Barrett WA, Fry DL (1980) Quantitative topography of experimental atherosclerosis. In: Nerem RM, Guyton JR (eds) Hemodynamics and the arterial wall. Proceedings from a specialists meeting. University of Houston, pp 94-99
20. Barret WA (1981) An iterative algorithm of multiple threshold detection. Proceedings of pattern recognition and image processing. IEEE, Computer Society, 273-278

Discussion

CARO: What was the duration of the study and did it include any time course information?

CORNHILL: The period of cholesterol feeding was for 8 months. No additional time points were taken.

BLACKSHEAR: Was the probability of developing sudanophilic lesions higher at the ductus scar? Is there any significant alteration in tethering or wall thickness at this point?

CORNHILL: The area at the ductus scar had some of the highest probabilities of developing atherosclerosis observed in the thoracic aorta of the cholesterol-fed pigs. There is certainly increased tethering and increased wall thickness at this point and other important parameters such as space available for arterial wall mass transport may differ at this unique anatomical location. Although this latter point has not been demonstrated experimentally, the region of the ductus scar is an area which has a very different structure from the rest of the wall.

S.M. SCHWARTZ: Isn't the information that you presented contradictory to most of the other things that we have been hearing, in that you said that flow was not important in the localization of lesion?

CORNHILL: The data which we presented is somewhat contradictory to the other data in that for what we see there is no good correlation with specific hemodynamic events (e.g. proximal/distal to the ostia) and the probability of developing atherosclerosis. At best we believe that the lateral leading edges of the flow divider have a high probability which may be associated with hemodynamic forces on the wall; however, this accounts for only a small percentage of all the lesions which were observed. There are many areas which have high probabilities of developing disease which are not areas where we would expect specific hemodynamic events to occur. They may be hemodynamic in the sense of arterial wall mass transport, but not in the sense of fluid blood flow affecting the arterial wall.

SCHWARTZ: Are you saying that endothelium is not important in atherogenesis?

CORNHILL: No, what I am saying is that, on the basis of these probability maps, we can now investigate the structural and metabolic state of the endothelium at specific locations of high and low probability of lesion formation and correlate these findings with other wall parameters.

STEHBENS: I think that it is a mistake if you just rely on sudanophilia to localize a lesion of atherosclerosis. I believe this to be of very limited value, particularly in man. The penetration of the sudan dye is restricted by both the degree and nature of intimal thickening, and you do not necessarily locate any of the sudanophilic material which may lie underneath. I think that we really ought to be looking at localization of intimal thickening because the overwhelming evidence suggests that lipid deposits in the vessel always occur initially in areas of intimal thickening. In any case, as you know, I do not regard lesions that you are producing as atherosclerotic anyway.

CORNHILL: Let me make a few comments. The mini-pigs used in this study were found to develop a classic humanoid atherosclerotic intimal disease including complicated plaques. Diet induced atherosclerosis in this model is associated with increases in the family of low density lipoproteins (e.g. LDL, IDL, HDL, VLDL and BVLDL) and a reciprocal decrease in high density lipoproteins (HDL_2). Thus, the pig model not only simulates the human disease structurally, but also it appears to reflect in exxagerated form acute human lipoprotein response to high cholesterol intake. It is of further interest to note that current studies in our laboratory show that spontaneously occurring athero-sclerotic disease in old (> 10 years old) normocholesterolemic mini-pigs has a topographic distribution and many histological features that are identical to those appearing in the diet-induced disease in younger members of the same species, thus lending firmer validity to the study of the mini-pig model.

Subject Index

MIX
Papier aus verantwortungsvollen Quellen
Paper from responsible sources
FSC® C105338

If you have any concerns about our products,
you can contact us on
ProductSafety@springernature.com

In case Publisher is established outside the EU,
the EU authorized representative is:
Springer Nature Customer Service Center GmbH
Europaplatz 3, 69115 Heidelberg, Germany

Printed by Libri Plureos GmbH
in Hamburg, Germany